ison

'Beauty in engineering is that which is simple, has no superfluous parts and serves exactly the purpose.'

HARRY FERGUSON

THE Ferguson TRACTOR STORY

STUART GIBBARD

Published by

Old Pond Publishing

First published 2000

reprinted 2002, 2009, 2013

ISBN 978-1-903366-08-0

A catalogue record for this book is available from the British Library

Published by

Old Pond Publishing Ltd
Dencora Business Centre
36 White House Road
Ipswich IP1 5LT
United Kingdom

www.oldpond.com

Frontispiece illustration:
One of hundreds of thousands of Ferguson tractors still working worldwide, this TE-20 is used for launching fishing boats on the Norfolk coast.

Cover design and book layout by Liz Whatling
Printed in China on behalf of
Latitude Press Limited

Contents

Acknowledgements 6
Author's Note 7
Preface 8
Foreword 9

Introduction — The Ferguson System 10
Chapter One — The Belfast Plough 16
Chapter Two — The Ferguson-Brown 28
Chapter Three — The Ford Ferguson 44
Chapter Four — The TE-20 Family 62
Chapter Five — Ferguson Overseas 94
Chapter Six — Massey-Harris-Ferguson 116
Epilogue — The Ferguson Legacy 147

Appendix 1 — Ferguson Conversions 151
Appendix 2 — Ferguson Tractor Serial Numbers and Designations 160
Bibliography 162
Index 163

Acknowledgements

The greatest enjoyment that comes from writing a book is not, as many will think, from seeing your name on the cover and your words in print. That is a novelty that quickly passes. No, for me, the real pleasure is in the research and the opportunities it brings - in particular, the opportunity to talk to the people who were in at the deep end. There is no greater privilege than being given the chance to meet, talk to and learn from the men who designed, built, demonstrated and sold the tractors. It is a pleasure that still remains undiminished after nine books.

Each book is different, as is each company, but the warm welcome, considerable assistance and encouragement that I receive from the people who patiently and knowledgeably answer my questions remain the same, and the Ferguson men were no exception.

These are the people who bring the story alive, not me. It is difficult not to be moved by their pioneering spirit and the inspiration that pervades their conversation, or be swept up by the emotion of the triumphs and tribulations as the events are recounted as if they were only yesterday. I often realise what a task I set these willing few when I expect them to remember every detail of something that happened over sixty years ago when I can't even remember where I put my pencil ten minutes ago.

There was, however, one difference between the Ferguson men and people I have interviewed from other organisations. I am always aware of an overwhelming sense of loyalty to the product or company, even if that concern has long since disappeared. But here the loyalty seemed almost tangibly greater, and much of it reserved for the one individual who was obviously a strong influence on their lives. That man of course was Harry Ferguson.

I have many ex-Ferguson personnel to acknowledge, but I would like to start by thanking Alex Patterson. Alex is not in the best of health, but his mind is as sharp as ever and his attention to detail is unsurpassed. His help was invaluable and there were few questions that he could not answer, having joined Harry Ferguson in 1938, and becoming superintendent of engineering during the TE-20 era. He also kindly read my manuscript and filled in any missing information.

I would also like to thank all the following ex-Ferguson men who have kindly made time for me, been at the end of a telephone, invited me into their homes or met me over a pint to share their wonderful reminiscences. They are: John Armstrong, Keith Base, Jack Bibby, Peter Boyd-Brent, Dick Dowdeswell, Bill Halford, Roy Harriman, Nigel Liney, Nibby Newbold, John Roberts, Colin Steventon, Peter Vernon and Bud White. To this list, I must add two Massey-Harris men, Brian Rogers and Jim Wallace, and Roy Bare from the Australian Ferguson dealership, R. H. Bare Pty Ltd.

I have reserved special thanks for Erik Fredriksen. Erik traced most of my contacts and set up the meetings with many of the above gentlemen. He kindly gave up three days of his time to accompany me (I could say navigate me, but maybe he should buy a better map) around the Coventry area on my visits and steer me around what few archives remain at Banner Lane. He helped me with a great deal of information, and for anybody wishing to learn more about the 'big Fergie', I can heartily recommend his recently published little booklet, *The Legendary LTX Tractor.*

I must pay special tribute to another ex-Ferguson man and author, Colin Fraser, whose acclaimed biography of Harry Ferguson has been a major source of information. *Harry Ferguson - Inventor & Pioneer* remains the benchmark Ferguson history and is every Ferguson enthusiast's bible. Colin's original research was both extensive and accurate, and I was delighted to be given the opportunity to meet both him and his lovely wife, Sonia, before my own book was completed.

Banner Lane is still home to the Massey Ferguson tractor, but the brand is now part of the AGCO Corporation, and I must thank that organisation and its personnel at the Coventry plant for their kind assistance in the preparation of this book. In particular, I would like to thank Jim Newbold and Ted Everett for kindly allowing me access to the photographic library.

Although John Briscoe was not directly involved in this book, he has been a helpful contact for all the years I have been writing. Thanks to him, I had already amassed a certain amount of information and photographs on Ferguson tractors before I even started this project. He often suggested that I should turn my attentions to grey or red/grey tractors rather than waste my time on other manufacturers while he patiently sourced photographs for my other books or articles. Ironically, by the time this Ferguson book was planned, he had retired from AGCO.

Ferguson tractors and equipment have a following like no other make, and the Ferguson enthusiasts and collectors who worship the grey are a learned and dedicated band, always willing to share their extensive knowledge of the marque. Without their help, this book would not have been possible. Five individuals in particular have given me unrivalled assistance, loaned valuable documentation and photographs, and allowed me access to their own research. They are: Selwyn Houghton, whose understanding of Ferguson-Brown tractors has no match; Mike Thorne, a true enthusiast who provided all the right contacts; Ian Halstead, who always said that I should write a Ferguson book and has been a source of continual help and information; Ben Serjeant, the man who began researching Ferguson history before anyone else and inspired me to follow his lead; and Jim Russell, whose Ferguson knowledge is eclipsed only by his photographic skill. Thank you gentlemen, it was most appreciated.

I must also thank Jim Russell and his, wife, Jane, for kindly providing me with (quality) accommodation and food during my time at Coventry, even if Jim did get his own back by keeping me up until after midnight with Ferguson bedtime stories. I am equally grateful to all the other Ferguson enthusiasts who have helped, provided information or photographs or allowed their tractors to be photographed, including Noel Collen, Colin Holwell, David Lory, Tom Lowther, John Moffitt, Greg McNeice, Evan Ould, John Popplewell and Robin Price.

I would also like to thank Mark Farmer, John Farnworth, John Foxwell, Peter Gascoigne, Phil Homer (Standard Motor Club), Sandra Nichols (Ford Motor Company), Bill Nunn, Stephen Perry (Perkins Engines), Derek Sansom (DSA), Clive Scattergood (Perkins Engines), Robin Shackleton (Hydro Agri), Bonnie Walworth (Ford Motor Company) and David Woods (New Holland).

Finally, after nine books, I think I should thank the following people for their patience and skills: my editor, Julanne Arnold; designer, Liz Whatling; publisher, Roger Smith (who also took several excellent photographs for the book); his (very long-suffering) wife and indexer, Lesley, and my (equally long-suffering) wife, Sue.

Author's Note

It is always a dilemma for an author of an historical book whether to use imperial or metric measurements. I was brought up with imperial units, but was forced into metrication during my teenage years. Being a traditionalist, I tend to favour the former.

The method I have tended to adopt for my other books is to use the system most in keeping with the period that I am writing about, and the era of the Ferguson remains firmly in imperial times. Having said that, however, most UK motor manufacturers preferred to quote engine capacities in cubic centimetres, while manufacturers in the USA used cubic inches. It seemed only right to do the same while giving the alternative figures in parentheses. Therefore, the sizes of the North American engines are given in cubic inches first followed by the metric equivalent in brackets and vice versa for the European power units.

The Standard Motor Company also tended to refer to its engines by their bore sizes in millimetres; for example, the 80 mm engine or the 85 mm engine. It would be illogical to present these any differently.

Several different systems for calculating or measuring engine power have been used over the years. Those figures that I can verify as true brake horsepower (bhp) are given as such, while others may be rated, drawbar or pto horsepower.

Preface

Writing a book about a machine as famous as the Ferguson tractor is a daunting proposition. Few tractors have captured the imagination as much as the 'little grey Fergie'. I can think of no other agricultural machine that has won such universal acclaim or been so fondly remembered. For many farmers, particularly those with smaller or more remote holdings, it was their first true taste of mechanised farming, and most people who have lived or worked in the country will have had some experience or knowledge of the Ferguson tractor.

The TE-20 has become an icon of Britain's engineering heritage of the twentieth century, as legendary as Sir Nigel Gresley's Pacific railway locomotives or Sir Alec Issigonis's Mini car. To the layman, the Ferguson is the classic tractor personified, and it still remains the first choice for many smallholders, 'weekend' farmers or the Pony Club fraternity with a small paddock to mow.

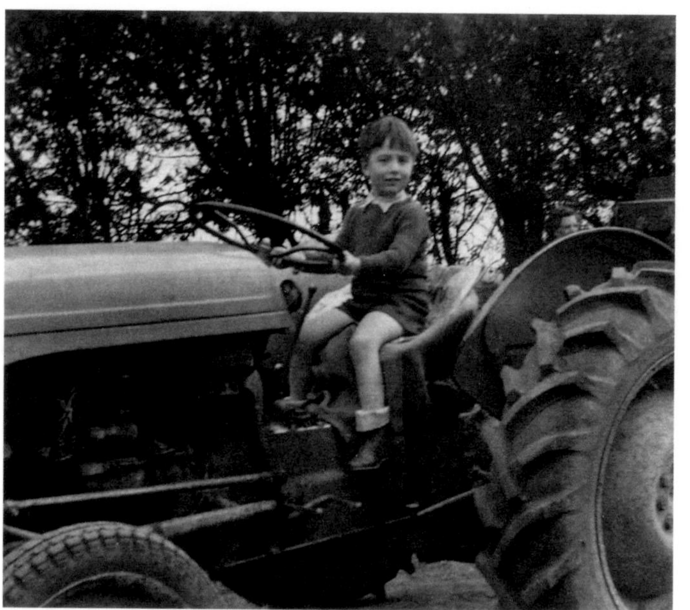

The author was introduced to the Ferguson System at a young age. This TE-20 model was the sole Ferguson among a fleet of Fordson tractors on his family farm.

Dare I say that having been brought up on an all-Ford farm, with just one solitary TE-20 that was used as a yard tractor and for spraying duties, I was brainwashed into believing that the praise heaped on the Ferguson was no more than just hype? We knew our Fordson Diesel Major tractors were more powerful and believed them to be superior, even if they lacked draft control hydraulics. A friend of mine once disparagingly remarked that the 'Fergie' was only really suitable for the man who had one acre and a goat. Had he got a valid point or was he being unfair to the Ferguson tractor?

An ex-Ferguson training instructor recently admitted to me that the tractor was possibly the weak link in the Ferguson System and had its limitations in terms of size and power. When towing from the drawbar, he said, 'It wouldn't pull the skin off a rice pudding.' His words, not mine.

However, he qualified this by saying that when full use was made of its hydraulic system with the correct implements, the Ferguson tractor was transformed, as if by magic, into an unbeatable machine that would run rings around the opposition. Here indeed was beauty in engineering.

It must not be forgotten that the Ferguson tractor brought affordable and practical mechanisation to the less-favoured areas and opened up the marginal land of many developing countries, or that it could till or plough hillside and stony land and inaccessible fields where no other machines could go. It brought motive power to upland and lowland farmers, market gardeners, hop, wine and fruit growers, local councils and municipal corporations the world over. Over one million Ferguson tractors were sold between 1936 and 1957, and over one million people cannot be that wrong.

I came to this story with an open mind, willing to be persuaded that the Ferguson tractor was or was not the stuff that legends are made of. No doubt, I have been more than a little influenced by the many ex-Ferguson men I have spoken to, and perhaps subconsciously affected by the magnetism of Harry Ferguson's words and philosophy that are the very fabric of every piece of Ferguson literature I have read. I was also greatly impressed by just how much more could have been achieved had the LTX prototype been allowed to go into production, and have finished this book a firm advocate of the Ferguson tractor.

Having said that, I have not embellished or glorified the story and have presented the trials, tribulations and mistakes as well as the triumphs, successes and achievements.

I have left it up to the reader to make his (or her - do I have any lady readers?) own mind up as to whether the Ferguson tractor really was the machine that revolutionised farming.

I have also tried to give both sides of the story of Harry Ferguson's successive partnerships that were formed on the road to realising his dream to build the tractor that finally replaced the horse. That vision is an inseparable part of the story and I felt I could not consider the tractor without looking at Ferguson, the man, and the greater concept of his complete farming system. This is not the first book to be written on Ferguson, nor will it be the last, but I hope it provides a further insight into the machine that became affectionately known as the 'little grey Fergie'.

STUART GIBBARD
July 2000

Foreword

By Alex Patterson, Engineering Workshop Manager, Harry Ferguson Ltd. and Massey Ferguson, 1945-1967.

My association with Harry Ferguson began in Belfast in 1938 when I joined his company as an apprentice. Little did I realise at the time that I would, in one way or the other, be connected with farm machinery and in particular with Massey Ferguson for the rest of my working life.

I always felt privileged to be part of the development of the Ferguson System and appreciated the importance of Mr. Ferguson's vision. In its most basic form this was a drive to improve the lot of the ordinary farmer in his quest to feed his family and indeed, on a grander scale, to help feed the world.

Harry Ferguson gave me the opportunity to contribute to the development of his farm machinery system and I learned so much from him. His key thoughts – simplicity of design and sound engineering principles – were always applied to his manufacturing and development philosophies. My association with Mr. Ferguson continued many years after he had severed his links with Massey Ferguson; we corresponded until the time of his death and I treasure his letters to this day.

Over the years I have been asked to contribute to many publications and documentaries regarding Harry Ferguson and I am more than happy to say that Stuart Gibbard's book is one of the most thoroughly researched and accurate accounts I have read.

ALEX PATTERSON
August 2000

INTRODUCTION
The Ferguson System

RIGHT: *Harry Ferguson, the inventor of the Ferguson tractor and the man responsible for pioneering the three-point linkage and a system of hydraulic depth control.*

'The machine that revolutionised farming' is a remarkable accolade, and one that has been applied many times to the Ferguson tractor. It is obviously a contentious statement as other machines, such as the combine harvester, have had an equal, if not arguably greater, impact on agriculture. But the importance of the Ferguson's contribution to the progress of mechanisation cannot be denied, and its principles are embodied in the design of almost every modern tractor or mounted implement. Its inventor Harry Ferguson claimed, 'The Ferguson System is an entirely new way of farming, born of a revolution in agricultural engineering and production methods'.

But what exactly was the Ferguson System? Most people are aware that the Ferguson tractor was famous for its hydraulic lift. It was not the first or only tractor of the time to have a power lift, but while other manufacturers did eventually develop their own basic hydraulic systems, none were as refined or as advanced as the Ferguson. What Harry Ferguson pioneered was a system of hydraulic depth control and an effective three-point linkage arrangement that allowed the tractor and implement to work together as one integrated unit. Although these are features we take for granted today as they have been adopted in some form by most of the world's major tractor manufacturers, we still need to consider the principles and benefits of the system to fully understand the development of the Ferguson tractor.

The agricultural tractors that first appeared on farms during the early years of the twentieth century were thought of as no more than towing vehicles to replace draught animals. They worked with adapted horse equipment that was trailed behind the tractor by a drawbar-hitch, pole or simply a length of chain. It was a very inadequate arrangement; the combination of the tractor and trailed implement made a cumbersome unit that could not be reversed into tight corners and needed a large headland for turning. Work-rates were not improved by the fact that most of the adapted farm machinery was designed to operate at speeds no greater than the clod-hopping gait of a heavy horse.

Because the trailed machines transferred little or none of their weight onto the tractor, the tractors themselves had to be built heavier with big driving wheels to increase traction. Soil compaction was becoming an issue, but there were other considerations; larger tractors meant less manoeuvrable tractors that were unnecessarily expensive to build and used more fuel. The implements also became heavier as they had to incorporate depth wheels and a self-lift mechanism.

Another problem was that the forces exerted by the trailed equipment in work tended to pull

LEFT:
The old way - towing the plough or implement by a length of chain. The plough was unable to transfer any of its weight on to the tractor and early machines, such as this Saunderson, had to rely on their heavy construction to provide the necessary traction.

down on the tractor's drawbar and lift its front wheels. This led to instability, particularly when working or ploughing uphill. Also, if the plough or implement hit a tree root or other hidden obstacle, the tractor was likely to rear up. In extreme cases, tractors were known to turn over with fatal consequences for the driver. The Fordson with its worm-drive rear-axle was particularly prone to overturning and several accidents were recorded.

Ferguson revolutionised tractor design by combining a lightweight tractor incorporating a three-point hitch with a range of light mounted implements. John Foxwell, the chief engineer of Ford Tractor Operations from 1964 to 1975, once wrote, 'The three-point hitch is probably one of the most under-rated innovations in this or any other century. It is to the farm tractor what a chuck is to the drilling machine.' It united the tractor with its implements and turned it from a prime mover into a self-propelled agricultural machine.

The three-point linkage arrangement consisted of a triangulation of hitch points, comprising two lower-link arms, attached low down to either side of the tractor's rear axle housing, with a single top-link position mounted in the centre above the housing. To this, Ferguson added a hydraulic system to raise and lower the link arms to lift the implement in and out of work.

The hydraulic pump was mounted in the drive line between the transmission and the rear axle. On the Ford Ferguson and TE-20 tractors, it was driven off the power take-off shaft. The pump sucked oil from the rear-axle housing and passed it through a simple control valve to a single-acting ram. The ram cylinder was housed internally above the rear-axle drive-gears. Its piston activated a connecting rod attached to a crank arm that turned a cross-shaft. Lift arms at each end of the cross-shaft were connected via lift-rods to the lower-links. The single-acting hydraulic system provided the power to raise the implement, which then relied on gravity and its own weight to lower itself back into work. The control valve, connected by an internal linkage to a single lever on a quadrant to the right of the driver's seat, operated the lift. Ball joints at the end of the lower-link arms and top-link allowed the implements to be quickly and easily attached.

The three-point linkage both pulled and carried the implement in work, transferring its weight onto the tractor's rear wheels. This meant that Ferguson could build a lighter, more economical and cheaper

BELOW:
The Ferguson way - the TE-20's three-point linkage both pulled and carried the implement in work, transferring its weight on to the tractor's rear wheels. The tractor and plough acted as a combined unit that was both light and manoeuvrable.

ABOVE:
Ferguson's converging three-point linkage as seen on the prototype 'Black' tractor that was built in 1933. It consisted of two lower-link arms and a single top-link to which the implement was attached.

BELOW:
An illustration of how the Ferguson System utilised the natural forces affecting the tractor and plough in work. The broken arrows show the weight of the implement being transferred on to the TE-20's rear wheels, while the solid arrow represents the forward thrust that held down the front wheels.

machine, making his tractors more manoeuvrable and more versatile. They could cultivate the most inaccessible corners of fields, work up against walls and hedges, and plough gardens and plots of land too small for even a horse to operate in. Moving from one field to another was also quicker as the implement was carried on the back of the tractor.

Another advantage of the system was that the line of pull of the implement at its working depth extended through the converging links of the three-point hitch to a theoretical point just behind the tractor's front axle. This was known as the 'virtual hitch point' and meant that the natural forces affecting the implement in work were converted into a strong forward and downward thrust that held the tractor's front wheels down. Consequently, the Ferguson was very stable, especially for hillside work, and was unlikely ever to overturn when used correctly.

Other manufacturers also developed simple hydraulic lift units, but Ferguson's design incorporated a unique and very important feature known as draft control, a system of automatically controlling the depth of the implement. While the latest modern systems boast electronic controls and a host of advanced functions, the first basic hydraulic systems relied on simple position or draft control mechanisms.

Position control allowed the lower-link arms to be either fully up or down, or set at any position in between. This meant that the implement's height was maintained relative to the tractor, but the system could not control the precise depth of the implement in the ground. With draft control, the link arms could only be fully up or down, and could not be maintained at any intermediate position. However, this system controlled the working depth of the implement in relation to the contours of the land by 'sensing' the changing draft present between the soil and the implement. The depth was governed by a sensing mechanism connected to the control valve, receiving signals either through the top-link (top-link sensing) or the lower-link arms (lower-link sensing).

Harry Ferguson can be credited with inventing draft control, and his tractors were the first to use the system. The Ferguson tractor had top-link sensing. The draft force exerted on the implement by the soil was re-directed through the top-link and absorbed by a heavy coil-spring that was connected by a linkage to the hydraulic control valve. When the force coincided with the position set by the driver using the quadrant lever, the valve closed holding the implement at the desired height. As changes in the land increased or decreased the draught force, the valve would open and close to regulate the working depth of the implement. When ploughing, these corrections would happen about once a second.

Apart from the obvious advantages of the plough or cultivator maintaining a uniform depth and the driver having finger-tip control over the implement, there were other benefits. The draft control system amplified the weight

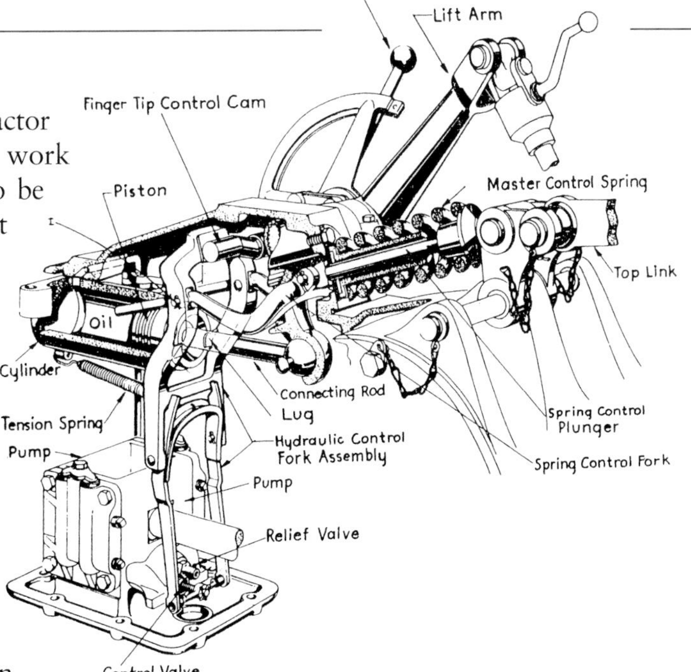

LEFT:
An exploded view of the TE-20 tractor lift mechanism. It shows all the main features of the Ferguson hydraulic system and its internal linkage.

transference, ensuring the weight of the tractor was automatically adjusted to suit the work or conditions. The implements could also be built lighter and to a simpler design without the need for extra depth wheels. Of great importance was the safety aspect: if the plough hit an obstacle, the system automatically released the weight of the implement, allowing the tractor's wheels to spin; the tractor came to a standstill and no damage was done. This led to Ferguson's linkage often being referred to as the safety hitch.

The simplest way to sum up the advantages of the system is to refer to the creed the Ferguson sales and field staff were taught by their instructors: 'The Ferguson System is a principle of linkage and hydraulic control over the depth at which the implement is working, enabling you to harness the natural forces created by the work that the implement was doing to add weight to the tractor.' Harry Ferguson's philosophy was that you did not need a sledgehammer to crack a nut, and he was proved right; not only did his system work, but it also changed the future of tractor development.

To take full advantage of the Ferguson tractor, you needed the correct implements to go behind it. To this end, Harry Ferguson introduced a full line of matched equipment that was eventually expanded to fulfil the needs of virtually every type of farming worldwide. In addition to the mounted implements, there were hydraulically operated tipping trailers and front-end loaders, and a complete range of accessories to make the most of your tractor.

LEFT:
The Ferguson hydraulic lift incorporated a system of draft control that allowed the plough to maintain a uniform depth. The single operating lever to the right of the seat gave the driver finger-tip control over the implement.

Ferguson's approach to farming went even further once the TE-20 was in production at Banner Lane. He set up an integrated dealer network, pioneered organised service in the field with a fleet of service vans and free on-farm visits, and established a mechanised farming school. Each implement and machine had its own beautifully presented instruction and parts book, and to help with finances there was the Ferguson 'pay-as-you-farm' hire-purchase plan. The Ferguson System was more than just a tractor with a three-point linkage and hydraulics – it was a complete vision of global mechanised farming. The result was an almost utopian view of agriculture that was reflected in the brochures and advertising material that were produced by Harry Ferguson's public relations man, Noel Newsome, a journalist who had been involved in propaganda work during the Second World War.

There were some disadvantages to the Ferguson System. Ferguson himself was fanatical about simplicity of construction – for example, one spanner had to fit all the nuts used on the tractors and implements and be all that was

RIGHT: *A typical Ferguson demonstration; the roped-off enclosure was used to show how it was impossible to work in such a small and inaccessible area with a horse and plough.*

required for all field adjustments. He insisted that only one lightweight size of tractor was needed; however, this was not enough for some of the larger acreage farms that required more power. Like the horses that his machines replaced, Ferguson sometimes seemed blinkered, resisting the demand for larger tractors and losing many sales in the process.

Another drawback to the tractor was that it could not be used with most of the existing implements that the farmer already had in his yard. If he switched to a Ferguson, then he had to face up to the extra expense of buying the matching equipment to suit it. However, this appeared to be no insurmountable obstacle as over half a million TE-20 models alone were sold.

Harry Ferguson's real success lay in his marketing skills; he managed to sell the farmers not just a tractor, but also a complete range of equipment and a totally new farming system. It was an achievement that has never been surpassed and is unlikely ever to be repeated. Colin Fraser, in his acclaimed biography of Ferguson, rightly calls him an inventor and pioneer, but he was equally a visionary and marketing genius.

The zeal with which Ferguson promoted his concepts made him seem more like a man on a crusade; a crusade to make a better world with safer farming methods, and in his later years, with safer motoring. His tractor was an educational tool to demonstrate a farming system that he earnestly believed everyone, from governments and industrialists to farmers and customers, should subscribe to. It has also been said that you never owned a Ferguson tractor; you paid for it, but it still belonged to Harry Ferguson. This comment arose because if Harry ever saw one of his tractors being misused in a field, he was likely to stop and give the driver a lecture!

Harry Ferguson, the man, was a complex character; a remarkable and gifted individual who displayed more than a trace of eccentricity. People who met him for the first time recall being surprised by his small stature and then overwhelmed by his intense nature and piercing blue eyes. He would be willing to talk for hours on his favourite topics of farming, machinery and engineering, and had a penchant for 'Irish' jokes; yet in private he was a shy man with little conversation outside his pet subjects.

BELOW: *The second part of the demonstration was designed to show how easily and quickly a young lad could cultivate the same plot with a Ferguson tractor. Dick Chambers, who ran the Ferguson School of Farm Mechanisation, normally allocated this task to his son.*

that Ferguson was able to draw on to turn his ideas into reality. There was also great poverty on the farms in the province, with large families existing on small acreages, and every last inch of land was used to grow crops. This need to cultivate right up to the hedgerow was the inspiration for the Ferguson System while Ulster provided the background to its development.

The size of the little Ferguson tractor was not only governed by the size of the farms in Ireland, but also by Harry's stature. He was a small man, and it is doubtful whether he could have coped with some of the larger and more cumbersome tractors on the market. The opinion is that Ferguson designed the tractor to suit his slight build, and it became the machine that even a boy could handle - like its inventor, a legend in its own lifetime.

LEFT:
The Ferguson System was a complete concept of mechanised farming. There was a full range of implements, each with its own beautifully presented instruction book.

BELOW:
Typical of the Ferguson sales literature of the period, this 1955 brochure for the tractors and implements extols the virtues of the Ferguson System.

Colleagues of Harry Ferguson have described his single-minded, intolerant and sometimes erratic personality. His ideals were perfection, flawlessness and precision, and he expected no less than one hundred per cent dedication from his team. But for all this, he was regarded as a fair-minded man, prepared to praise accomplishment and give encouragement when necessary; he inspired great loyalty and those who worked for him remember 'HF' with deep affection.

Just how much of Ferguson's accomplishments were more attributable to his engineers, including Willie Sands, Archie Greer and John Chambers, has always been a subject of controversy. However, Harry Ferguson provided the motivation and was the catalyst that brought the Ferguson System to the world; the success of the tractors and implements was undoubtedly the result of his vision, energy and enthusiasm.

Alex Patterson, a fellow Ulsterman who worked with Harry Ferguson longer than most, firmly believes that the concept embodied in the Ferguson tractor could not have been developed by anyone else, or would have happened in any other country than Ireland.

Northern Ireland had important and established shipbuilding, linen and tobacco industries that nurtured the engineering skills

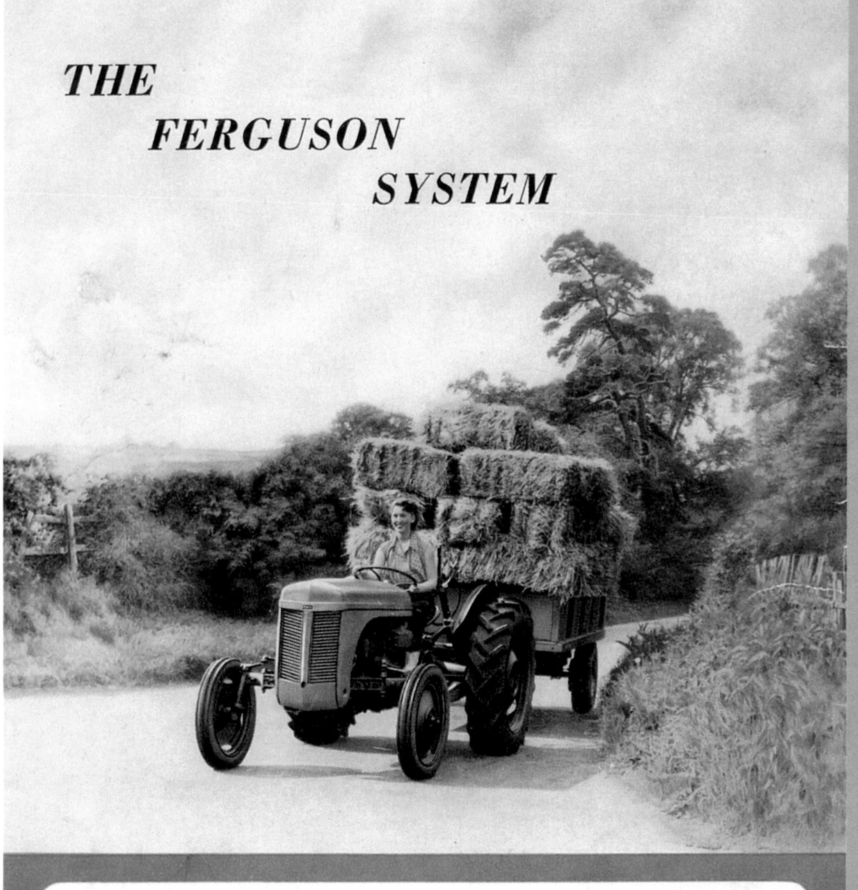
THE
FERGUSON
SYSTEM
WHY IT HELPS TO CUT YOUR
FARMING COSTS AND RAISE
YOUR OUTPUT AND INCOME

CHAPTER 1

The Belfast Plough

Willie Sands demonstrates the Duplex plough behind a Model F Fordson tractor in 1921. Harry Ferguson is on the far left of the group of spectators and is believed to be standing next to Thomas McGregor Greer. The location has been suggested as Bullock's Farm, just outside Belfast.

RIGHT: *Harry Ferguson's birthplace, Lake House at Growell, near Dromore in County Down.*

There are not enough houses in Growell for it even to be called a hamlet. It is no more than a small farming community near Dromore, about sixteen miles from Northern Ireland's capital city of Belfast. Pretty enough when the sun shines, but more often grey and damp, the place is unremarkable and would remain insignificant were it not for the fact that it is the birthplace of one of Ulster's most famous sons.

Harry Ferguson came from a strict Protestant family of Scottish descent who had farmed in County Down for several generations. His parents, James and Mary Ferguson, lived at Lake House in Growell. They had eleven children; their fourth child, born on 4 November 1884, was christened Henry George but was always known as Harry.

Although they owned their own farm, a little over 100 acres of mainly grazing land, the Fergusons were by no means wealthy; farming in Ulster at the turn of the century was a struggle, and all the members of the family were expected to help out. James Ferguson's deeply held religious beliefs made him a stern disciplinarian, and even the youngest children were given chores to do on the farm.

Harry left school at the age of fourteen and worked on the farm for another four years. He was not really suited to the job and did not enjoy farming; he was not as physically strong as his seven brothers and had an aversion to the tedium and drudgery of farm work, preferring to pass his time experimenting with things mechanical. In 1902 he left the farm and became an apprentice mechanic, working for his eldest brother, Joe, who had set up a small repair shop for motor cars and cycles in the Shankill Road area of Belfast.

Belfast was becoming a

BELOW: *The small farming community of Growell in Northern Ireland. Ferguson's family had farmed in this area for several generations.*

LEFT:
Harry Ferguson - the aviator. He built his first aeroplane in 1908 and succeeded in making the first powered flight in Ireland on 31 December 1909.

great shipbuilding and industrial city, and was regarded as the transportation and commercial centre of Ulster. As the city prospered, so did Joe Ferguson's workshop due to the increasing number of cars on the road. Harry found he had a natural aptitude for tuning engines, and J. B. Ferguson & Co. soon built up an enviable reputation for quality work.

Even though he was still little more than a callow youth, Harry Ferguson was already showing a remarkable flair for marketing and inventiveness. From 1904, with the business financially secure, he was able to demonstrate both his showmanship and the wilder side of his nature by entering a number of car and motorcycle races to promote his brother's company.

ABOVE:
The occasion is not recorded, but the Ferguson family look ready for a day out with Harry at the wheel of a Vauxhall from May Street Motors. Harry Ferguson opened this garage in Belfast in 1911, taking on several agencies including Vauxhall, Darracq and Maxwell.

Taking his passion for dangerous sports one step further, he began building his own aeroplane in 1908. Designed as a monoplane, it was fitted with a V8 engine supplied by J. A. Prestwich of London. This JAP engine was air-cooled and developed 35 hp. On 31 December 1909, the aircraft flew 130 yards; Ferguson had made the first powered flight in Ireland, and had become the first British aviator to build and fly his own aeroplane. It was said at the time that the aeroplane might have gone further had Harry been a better pilot, but the truth was that he was very short-sighted and could not see where he was once he got more than a hundred feet off the ground. Further exploits (and crashes) with the aeroplane and racing cars brought Ferguson fame, but led to disagreements with Joe who regarded Harry's reckless stunts as a waste of time and money. There was also a little ill-feeling as it appeared that both brothers were courting the same girl, Maureen Watson, a grocer's daughter from Dromore.

In 1911, Harry left to set up his own garage in May Street, Belfast, with financial support from two of his friends: Thomas McGregor Greer, a wealthy landowner with a passion for cars, and John Williams, whom Ferguson had met while

ABOVE:
During the First World War, Harry Ferguson took on the agency for the American Overtime tractors. They were well-built machines and the Model N shown was rated at 25 hp.

attending evening classes at Belfast Technical College. The garage was called May Street Motors and it both repaired and sold cars, and Ferguson became the agent for several leading British, American and French makes. During 1912, he changed the name of the company to Harry Ferguson Ltd. and his success was assured as customers were drawn in by the notoriety of his flying and racing escapades. With his business on a firm financial footing, he felt ready to propose to Maureen Watson, who had become a stabilising influence on his life. His proposal was accepted and they married the following year.

Ferguson passionately believed that mechanical

RIGHT:
Harry Ferguson discovered many drawbacks in the design of the agricultural tractor during his 1917 tour of Ireland to promote better ploughing techniques for the Irish Board of Agriculture. He believed the practice of using a heavy tractor, such as this International Titan, to pull a cumbersome plough by a length of chain was both inefficient and crude.

LEFT:
Ferguson's first plough was introduced in 1917 and cost £28. Often referred to as the 'Belfast' plough, it was designed to fit the Eros, a tractor conversion of the Model T Ford car.

developments should be accessible to everyone and not just the few who could afford them, and should be linked to some improvement in the quality of life. It seems that he and his brother, Joe, collaborated on a project to build a 'people's car' that needed minimum maintenance, with routine servicing at only six-month intervals instead of every week as was usual at that time. The car, known as the Fergus and based on Vauxhall running gear, was introduced in 1915, but never really caught on.

The First World War saw the British Government campaign for increased production to help beat the U-boat blockade. The call to arms had seriously deprived farms of men and horses, and Ireland was as much affected as mainland

LEFT:
The Eros as used by Ferguson to demonstrate his new plough was an American conversion marketed in Britain by Morris Russell of London. This replica machine is based on a British Pattisson conversion made in Middlesex with Model A Ford running gear.

RIGHT:
The first design of Ferguson plough for the Model F Fordson tractor was based on the original 'Belfast' plough and was simply hitched to the tractor's drawbar cap.

BELOW:
The improved Ferguson plough for the Fordson appeared in 1919. It was attached to the tractor by two parallel drawbars known as the Duplex hitch.

Britain. Greater mechanisation was needed, and Harry Ferguson took on the agency for the American Waterloo Boy tractors that were sold in the UK under the Overtime name. Ably assisted by Willie Sands, a brilliant young mechanic who had joined May Street Motors in 1911 after serving an apprenticeship in the linen industry, he organised a series of demonstrations of the Overtime working with a Canadian three-furrow Cockshutt plough.

Ferguson and Sands staged the demonstrations with such alacrity and precision that they soon came to the attention of the Irish Board of Agriculture. The Board asked Ferguson to tour Ireland on its behalf during the 1917 spring ploughing campaign, instructing farmers on the

correct operation of their tractors and ploughs to increase their performance and standard of work. It was to prove a pivotal time for Ferguson; his experiences with these basic agricultural machines highlighted their obvious failings and made him question their efficiency. He realised that there must be better ways of connecting the plough to the tractor other than dragging it along by a length of chain. The implement, he maintained, must become an integral part of the tractor, and he began designing his own plough.

Having heard that there were plans afoot to build the proposed new American Fordson tractor at Cork in Ireland, Harry Ferguson engineered a meeting with Henry Ford's right-hand man, Charles Sorensen, in the early summer of 1917. Sorensen was in London setting up a contract to supply tractors to the Ministry of Munitions. He listened to Ferguson's plans for an integral plough and agreed to consider his ideas and pass them on to Henry Ford.

The first Ferguson plough, often referred to as the 'Belfast' plough, appeared in the December of that same year, and was the result of a combination of Harry Ferguson's inventiveness tempered by Willie Sands's engineering ability. It was designed for use with the Eros, a tractor conversion of the Model T Ford car.

Built by the E. G. Staude Manufacturing Company of Minnesota and marketed in Britain by Morris Russell & Co. of London, the Eros conversion provided a lighter and inexpensive alternative to many of the larger tractors on the market. Ferguson's plough was hitched forward of the tractor's rear axle and lifted by an arrangement of levers and balance springs. It was advertised for £28 and was shown throughout Ireland during 1918, culminating in a demonstration at Chelmsford in England in December. The Eros plough was not entirely successful, but it generated enough interest to persuade Ferguson to press on with plans for a revised version to suit the new Model F Fordson.

The Fordson tractor appealed to Ferguson because it was simple, compact and reliable, and Henry Ford's mass-production techniques meant that it was very competitively priced. The first Ferguson plough for the Fordson was no more than an adaptation of the Eros design and was simply hitched to the tractor's drawbar cap. It was a cumbersome arrangement and the plough was still relatively crude and heavy.

Taking the design one stage further,

ABOVE:
The Duplex plough at work. It was only about one-third the weight of similar implements of the time and set the design for the modern tractor plough. Control levers adjusted depth and altered the furrow width.

The "Ferguson" Plough

The only criticism that has been advanced against our Plough is that we are not an old established firm of Plough Manufacturers. To this we reply, that to have been first merely proves Antiquity, but to have become first proves Merit.

Manufactured by
HARRY FERGUSON LTD
May Street, BELFAST.

W. & G. BAIRD, LTD., BELFAST.

LEFT:
One of the earliest pieces of Ferguson literature, this brochure for the Duplex plough was issued in 1922. The statement on the front cover is a typical Harry Ferguson sentiment.

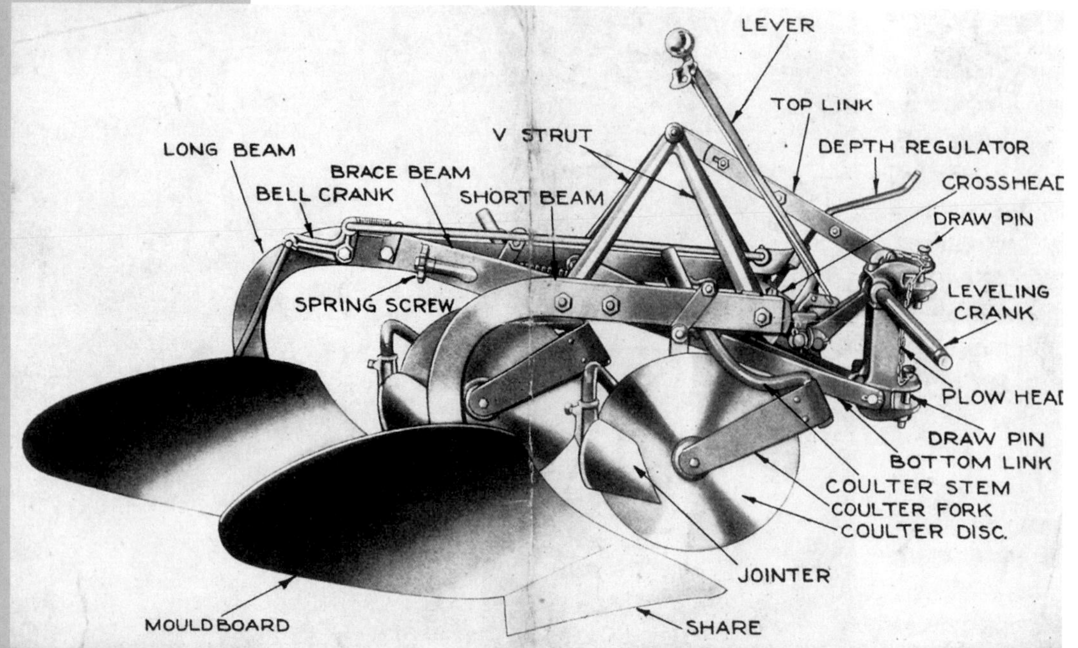

ABOVE: *The Ferguson-Sherman plough as built in America from 1925. The large lever was assisted by a balance spring to lift the plough out of work. The bell-crank was attached to a floating skid or 'slipper' that regulated depth.*

to the tractor. A mechanical lift arrangement, operated by a lever and assisted by a balance spring, lifted the plough in and out of work. The hitch and plough were ready by 1919 when the Irish production of the Fordson began at Ford's new plant in Cork.

The following year, Ferguson took Sands to the United States in an attempt to persuade Henry Ford to manufacture the plough in America. It proved to be an unsatisfactory meeting. Ford misread the Irishman's intentions and merely offered him employment. When this was declined, he tried to buy the plough patents. Ferguson politely refused the offer and returned to Belfast, whereupon the fellow directors of Harry Ferguson Ltd. suggested he give up the costly plough experiments.

But it was not in Harry Ferguson's nature to give up; he stubbornly carried on refining the plough, making modifications and improving the design. A disillusioned Willie Sands had left to set up his own business, and his place was taken by a pattern maker, Archie Greer. Tests were

Ferguson and Sands then devised the Duplex hitch with two parallel clevis-drawbars, each hitched to a horizontal plate with five holes for adjusting furrow width. The upper plate was mounted to the tractor's transmission housing, while the lower plate took the form of a casting that replaced the Fordson's original drawbar cap.

The plough itself was lighter and had a single top link and two bottom links connected to a headstock. The headstock incorporated the Duplex hitch and was attached by two draw pins

RIGHT: *The Ferguson-Sherman plough behind a Model F Fordson, possibly in Ireland. The tractor has the long-wing fenders that were designed to stop it overturning.*

conducted in conjunction with the Ford factory at Cork, and much of the field work was carried out by Patrick Hennessy who later became the head of the Ford Motor Company in Britain. In 1922, Ferguson produced a brochure for the plough and sent a circular to various machinery dealers inviting them to a demonstration to be held on 27 and 28 March under the auspices of the Ministry of Agriculture at the Farm Institute near Sparsholt in Winchester.

Sadly, the plough still had its limitations and was not a commercial success. Sands returned to the fold and devised a floating skid that was mounted to the rear of the plough to improve its depth control, but it was not enough to revive interest in Britain. After one or two abortive attempts to get the plough produced in the United States, Ferguson managed to reach an agreement with the brothers Eber and George Sherman, who were the Fordson tractor distributors for the state of New York. Manufacturing facilities were set up in Evansville, Indiana, and a new company, Ferguson-Sherman Incorporated, was established in December 1925.

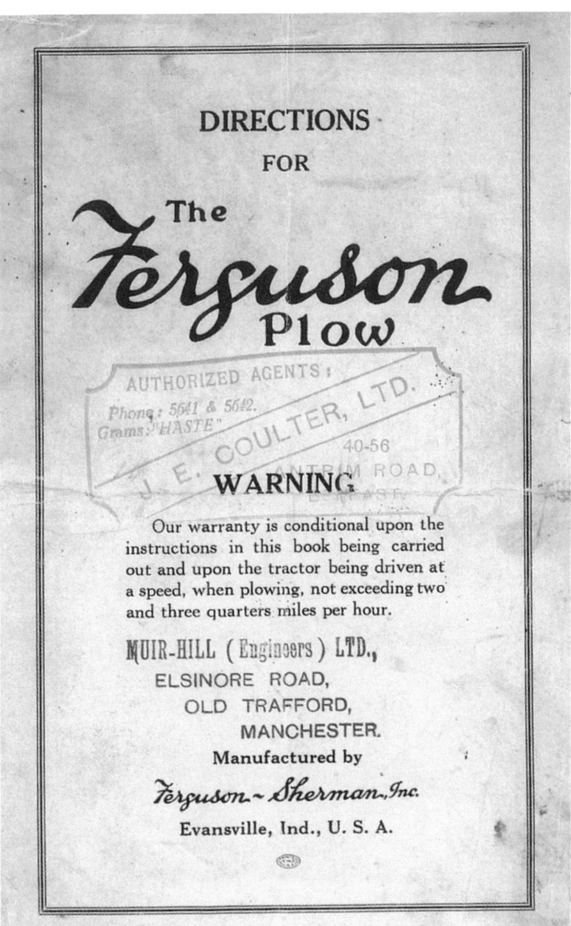

DIRECTIONS
FOR
The
Ferguson
Plow

AUTHORIZED AGENTS:
Phone: 5641 & 5642.
Grams: "HASTE"
J. E. COULTER, LTD.
40-56
ANTRIM ROAD,
BELFAST.

WARNING

Our warranty is conditional upon the instructions in this book being carried out and upon the tractor being driven at a speed, when plowing, not exceeding two and three quarters miles per hour.

MUIR-HILL (Engineers) LTD.,
ELSINORE ROAD,
OLD TRAFFORD,
MANCHESTER.

Manufactured by
Ferguson-Sherman, Inc.
Evansville, Ind., U. S. A.

LEFT:
The cover of the instruction book for the Ferguson-Sherman plough. Note it bears the stamp of both the British importers, Muir-Hill (Engineers) Ltd., and the agents in Ulster, J. E. Coulter Ltd. of Belfast.

BELOW:
Ferguson's designs for a three-point hitch were finalised in 1928. The linkage is seen fitted to a 1925 Model F Fordson. The arrangement incorporated a hydraulic lift and a draft control system with lower-link sensing.

Several of the American Ferguson-Sherman ploughs were brought into Britain by Muir-Hill Service Equipment Ltd., later Muir-Hill (Engineers) Ltd. of Old Trafford, Manchester. Walter Hill, the son of one of Muir-Hill's founders, had joined the British Ford Motor Company, based in nearby Trafford Park. During a spell working for Ford in Detroit, he arranged for Muir-Hill to import American attachments for Fordson tractors, including the Ferguson-Sherman ploughs that were offered in the UK until 1928.

With some returns coming back on his investment, Ferguson with his small band of Sands and Greer set about improving the way the plough was hitched to the tractor and developing a better system of depth control. After due consideration, they hit upon the idea of using the draft forces

ABOVE: *The motor business of Harry Ferguson Ltd. opened its new Belfast showroom in Donegall Square in 1928 and was appointed the Northern Ireland distributor for Austin cars. This later photograph of the premises dates from about 1947.*

affecting the implement to regulate its working depth automatically. It was a new concept, and Ferguson decided to patent the principle. This landmark patent, entitled 'Apparatus for Coupling Agricultural Implements to Tractors and Automatically Regulating the Depth of Work', was granted in June 1926.

After a great deal of experimentation, Ferguson's team arrived at a system of depth control that worked by mechanically sensing the torque in the tractor's rear axle, relying on the worm gear in the Fordson final-drive to lift or lower the plough. This was followed by plans for a hydraulically operated three-point hitch arrangement with converging linkage – the forerunner of all modern systems.

The design was finalised in 1928, and was fitted to a 1925 Model F Fordson. A four-cylinder piston pump powered a single vertical ram to operate the linkage. The arrangement incorporated lower-link sensing and was the first Ferguson hydraulic draft control system.

By 1928, the motor business, Harry Ferguson Ltd., was flourishing; it had become the Northern Ireland distributor for Austin cars and had opened a new showroom in Donegall Square East, Belfast. Things were not so rosy for the plough enterprise as that same year Ferguson's American interests were badly hit by the Ford Motor Company's decision to end production of the Model F Fordson. Another suitable tractor had to be found, and Ferguson approached other manufacturers with his system. Allis-Chalmers in the USA and Ransomes & Rapier of Ipswich showed interest

LEFT:
The Ferguson plough attached to a prototype Rover tractor for trials in 1931. It was an unsuccessful combination as the tractor was unstable in the furrow and any plans to put it into production were dropped.

but were reluctant to commit themselves to the venture. A deal was nearly struck with Morris Motors, but the Oxford-based company got cold feet and pulled out at the last minute.

Another British car manufacturer, the Rover Company of Coventry, also had plans to build a tractor and indicated an interest in the Ferguson linkage. A prototype was built in 1931 at Braunston Hall, home to Rover's managing director, Colonel Frank Searle. It was powered by an 839cc twin-cylinder air-cooled engine designed for the Rover Scarab car and was fitted with Roadless tracks. Lieutenant-Colonel Philip Johnson of Roadless Traction was a former army colleague of Searle's and was involved in the design. The tractor was equipped with a Ferguson plough, but was found to be too unstable in the furrow and was scrapped at the end of the year.

Although the Fordson was back in limited production at Cork, few were exported to the USA and the Ferguson-Sherman venture had collapsed. Harry Ferguson was running out of options, and realised that the only opportunity he would have to exploit his system was to design and build his own tractor.

CHAPTER 2

The Ferguson-Brown

An early Ferguson Type A tractor, possibly a pre-production model on trial in 1936. The driver is believed to be Bob Annat, a trusted demonstrator who ran the training school for Ferguson-Brown equipment.

Harry Ferguson had great respect for the Fordson and admired its uncomplicated design and simple unit construction with the engine, gearbox and rear transmission forming the backbone of the transmission. The first Ferguson tractor was very much a scaled-down version of the Fordson, utilising the concept of unit construction, but built much lighter, relying on the principles of the draft control system and three-point linkage to provide traction.

Plans for the Ferguson tractor were drawn up during 1932. The small design team of Ferguson, Sands and Greer was joined by John Chambers, the son of a Downpatrick farmer. Chambers had served his apprenticeship with the famous Belfast shipbuilders, Harland & Wolff, who had built the ill-fated Titanic liner. He had planned to become a ship's engineer, but a propensity for seasickness forced him into a change of career.

The calculations for the strength of materials and several of the drawings for the tractor were made by John Anderson, an engineering lecturer at Lisburn College. As the design began to take shape, components were obtained from a number of outside suppliers, including a transmission, rear axle and steering box from the renowned gear manufacturers, David Brown & Sons Ltd. of Huddersfield in Yorkshire. Patterns and castings were sourced locally from Harland & Wolff and the seaplane manufacturers, Short Brothers.

The engine was an American unit supplied by the Hercules Corporation of Canton, Ohio, and was part of its IX series only introduced that year. Hercules had built the original engines for the

BELOW: *Harry Ferguson's prototype 'Black Tractor'. Completed in 1933, it was powered by a Hercules IX B engine and rated at 18 hp. Photographed at Stoneleigh, it is now preserved in the Science Museum in London.*

LEFT:
The 'Black Tractor' in the company of some later Ferguson machines in the engineering department at Fletchampstead after it had been refurbished in about 1950. The large tractor seen second from the right in the background is a prototype Ford-Ferguson 4P model.

Fordson, and this probably influenced Ferguson's choice of power unit, coupled with his belief that the Ferguson tractor's main market would eventually be the USA. The IX series was a lightweight range of four-cylinder side-valve petrol units with a stroke of 4 in. The IX A model had a 3 in. bore, while the IX B engine chosen for the Ferguson had a 3¼ in. bore giving it a capacity of 2,175 cc (133 cu in.) and was rated at 18 hp. In the USA, the IX A engine was fitted to Plymouth and Silver King tractors built by the Fate-Root-Heath Company, and was later used in Cleveland Cletrac machines, while the IX B powered Parrett and Kaywood tractors.

The prototype Ferguson tractor was assembled in the workshop of the May Street premises. Many of the parts were hand-made by the small team with Sands, Chambers and Greer working as designers, engineers and fitters. Several of the castings were made from aluminium alloy to keep the weight of the machine as light as possible. The tractor was ready by 1933, and was painted black at Ferguson's suggestion to reflect its simplicity.

Dubbed the Ferguson 'Black Tractor', it had a three-forward and single-reverse-speed gearbox. Unlike the Fordson that used a worm-drive in the rear axle, the Ferguson had a spiral-bevel crown wheel and pinion. Internal-expanding brakes fitted to the rear hubs were independently operated by two foot-pedals to aid turning. Naturally, the tractor had three-point linkage and

BELOW:
An early Ferguson tractor working with a two-furrow plough. Although very similar to the prototype 'Black Tractor', the Type A had an improved hydraulic system and revised linkage.

RIGHT:
A Ferguson Type A tractor in May 1936 at Dormington Court Farm near Hereford. Note the early features including the Coventry Climax engine and the Donaldson dry air-filter. The 6 x 4 yd roped-off enclosure behind the tractor was used to show its ability to work in limited spaces and cultivate up to walls and hedges.

BELOW:
Harry Ferguson takes the wheel of a Type A tractor during the Dormington Court demonstration.

incorporated a version of the hydraulic draft control system, powered by a four-cylinder plunger-type pump with an aluminium body and cast iron pistons and cylinders. The pump was driven off an extension of the gearbox layshaft, which meant that the hydraulics only worked when the tractor was moving.

Field trials showed up some inadequacies of the hydraulics, but further experimentation led to the adoption of top-link sensing that greatly improved the system. Other refinements included moving the control valve from the pressure to the suction side of the pump.

Believing that his tractor was nearly ready for production, Ferguson arranged a series of demonstrations, first in Ulster, then on land close to the David Brown works in Huddersfield, and finally at Ascot in Berkshire. Interest was shown by the Craven Wagon and Carriage Works, a subsidiary of the Sheffield steel company, Thomas Firth and John Brown, which eventually entered into an agreement to manufacture the tractor. It was a short-lived venture; the Sheffield company found Ferguson difficult to deal with, and were more than pleased when in 1935 David Brown stepped in and offered to take over the manufacturing agreement.

David Brown & Sons Ltd. was a long-established and well-respected engineering concern. David Brown had been a

ABOVE:
The Type A tractor was better known as the Ferguson-Brown. It was built at David Brown's Park Works near Huddersfield and production began in June 1936.

LEFT:
The Coventry Climax E Type engines that powered the early Ferguson-Brown tractors were based on the Hercules IX B unit and were probably made under licence from the American company. The similarity between the power units is emphasised by the fact that this Coventry Climax engine, shown with the head removed, was fitted with Hercules valves.

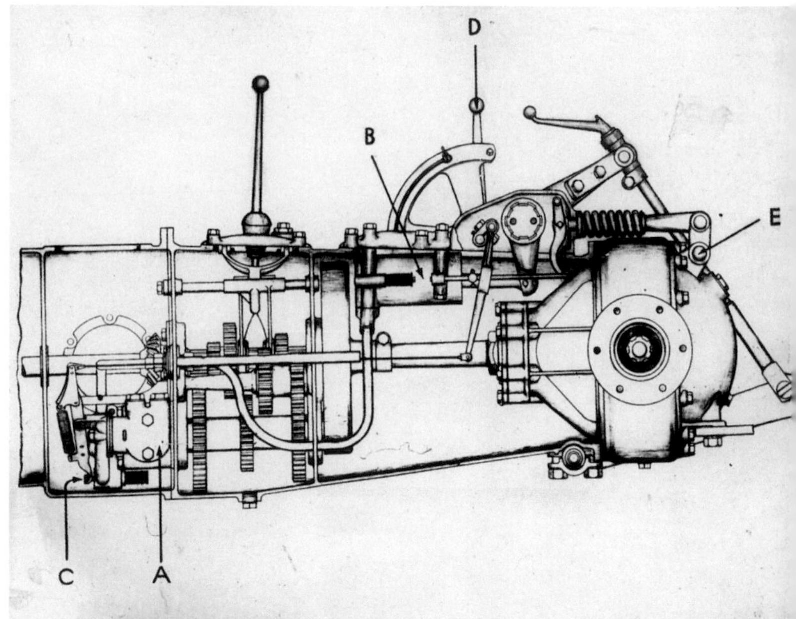

RIGHT:
A diagrammatic view of the Ferguson-Brown tractor's hydraulic mechanism. The hydraulic pump (A) fed the ram cylinder (B). The oil to the cylinder was controlled by a valve (C) that was operated by the lever (D). Draft signals sensed through the top-link position (E) also caused the valve (C) to open or close to maintain the correct depth of the implement.

THIS LEVER HAS A PAGE TO ITSELF. IT IS THE MOST IMPORTANT THING IN POWER FARMING
It controls and maintains an even depth of cut under all conditions.
It carries out your wishes automatically and instantaneously.
It takes the labour out of farm work, and is so simple that a child can operate it.
Impossible you say
CHALLENGE US TO PROVE IT

RIGHT:
Harry Ferguson Ltd. sales literature extolling the virtues of the Type A tractor's hydraulic system.

pattern maker who set up in business in 1860. After his three sons, Ernest, Frank and Percy, joined the firm, the company expanded into gear manufacture, moving into a new factory known as Park Works at Lockwood, near Huddersfield, in 1902. By the 1930s, the company had diversified into worm gears and vehicle transmissions. Frank Brown was now the chairman, and his son, David, was appointed managing director at the age of twenty-nine in 1932.

Young David Brown, who had joined the company as an apprentice in 1921, was a dynamic personality who had his own ideas for developing the business. He opened a new foundry at nearby Penistone in 1935, and was keen to move the company in new directions. He attended one of Ferguson's demonstrations, was impressed with the tractor and decided he would like to build it.

David Brown's agreement to manufacture the Ferguson tractor was made against the wishes of his father, Frank, who could see no future in the project. Undeterred, he set up David Brown Tractors Ltd., and managed to persuade the parent company to take a small financial interest in the concern and rent him space to build the tractors in a disused assembly shop at the Park Works.

Under the terms of the agreement, David Brown Tractors Ltd. was only the manufacturing concern. Sales, marketing and distribution, as well as design and engineering, were handled by a separate company, trading as Harry Ferguson Ltd., operating out of offices vacated by Karrier Motors in Cable Street at Longroyd Bridge, just two miles from Park Works. Harry Ferguson moved to Yorkshire to head his company and rented a house called Dungarth in Honley on the outskirts of Huddersfield.

Ferguson remained characteristically autocratic in his organisation with a triumvirate of three key personnel under him. Sands was his chief engineer, while Greer was chief draughtsman. Another apprentice from Belfast, Trevor Knox, was brought over from Ireland to act as chief demonstrator and later became sales manager.

LEFT:
Ferguson Type A serial No.1. Harry Ferguson promised this first production tractor to Thomas McGregor Greer and it was eventually delivered to his Tullylagen estate in January 1937. It has many later features, including pneumatic tyres, the oil-bath air-cleaner and non-standard fenders. The tractor is seen at Banner Lane, where it was put on display at the Massey Ferguson museum.

LEFT:
The main assembly area for the Ferguson-Brown was on the first floor of the tractor shop at David Brown's Park Works. The main components, such as the hydraulic units and transmission, were put together at one of five separate stations that can be seen behind the final assembly line.

RIGHT:
The second floor of the tractor shop at Park Works. This is where the implements were assembled and the tractors were stored prior to despatch. Note the plough bodies, beams and legs awaiting assembly by hand.

Seven or eight pre-production tractors were assembled in the spring of 1936. At least one of these, No.3, was used as an experimental model and was fitted with a fabricated steel transmission housing. Its gearbox was revised with the main and auxiliary shafts switched around, the gear lever offset and spur-gears added to provide a power take-off drive. This version was never put into production, and only the one example of this type has been recorded.

The production tractor was painted battleship grey; Ferguson had agreed that black was a little too austere, but wanted a colour that would blend into the countryside. It was officially called the Ferguson Type A, but was more often known as the Ferguson-Brown, although Harry Ferguson

RIGHT:
An early Ferguson Type A tractor harvesting with a binder.

retrospectively referred to it as the 'Irish tractor'. Built using heat-treated high-tensile alloy steel, with the barrel housing for the clutch and hydraulic pump and the transmission housing cast from aluminium alloy, the Type A weighed only 16½ cwt. There were only slight differences between the production machine and the prototype 'Black Tractor', although revisions had been made to the hydraulic system and linkage, and the clutch pedal had moved from the right to the left side of the barrel housing. The main change was that the engine was now a British-built Coventry Climax unit developing 20 hp.

Coventry Climax Engines Ltd. of Widdrington Road, Coventry, built a range of engines for industrial and automotive applications. The unit supplied for the Ferguson-Brown was known as the E Type engine. It was very similar to the Hercules lX B power unit, except that it used a British Solex carburettor, and was probably built under licence from the American company. Some Coventry Climax engines were even fitted with Hercules valves. Apart from the Ferguson tractor, the E Type engine was also fitted to GEC generating sets.

Although mention was made in the early press releases of the Ferguson-Brown being able to run on paraffin or vaporising oil, it was initially offered only as a petrol tractor. It did have a twin-compartment fuel tank, but the smaller compartment was only a two-gallon reserve for the main eight-gallon tank. Harry Ferguson was very much against the use of a heavier distillate fuel as it went against his ideals of clean simplicity. Although it was cheaper, it was not as efficient and needed a complicated vaporising system to burn it. There were also some problems with the sump oil being diluted by unburnt fuel. Ferguson's

ABOVE:
A Ferguson-Brown tractor with the Coventry Climax engine. Around 250 machines were built with this power unit before a changeover to David Brown engines began during 1937. Pneumatic tyres became optional the same year.

LEFT:
A Ferguson-Brown tractor working in a cabbage field with a rowcrop cultivator. Narrow 6 in. rear steel wheels for rowcrop work were introduced during 1937.

BELOW:
An advertisement for the Ferguson-Brown aimed at the export trade. The initial demand for the tractor was disappointing and very few were sold abroad.

booklet *Hints for Salesmen*, written in 1937, stated, 'Ninety per cent of the troubles experienced with tractors can definitely be traced to the misuse of paraffin.'

The choice is Yours!
HORSES
picturesque
but
out-of-date
the 'FERGUSON'
efficient
and
profitable
The profit making potentialities of Ferguson Farm Machinery (with its automatic depth control and patent hitch which makes the implement track with the FRONT Wheels) enables traders to offer, not merely a tractor but mechanism which entirely supersedes horses.
Overseas Agriculturists and traders are cordially invited to visit the Ferguson-Brown Works.
Nous invitons cordialement tous les agronomes et tous les commerçants à l'étranger à visiter les Usines de Ferguson-Brown.
Wir laden jeden Landwirt und jeden Kaufmann im Ausland ein, die Werke von Ferguson-Brown zu besuchen.

The first press announcements for what was heralded as a 'new British tractor' appeared in March 1936, and in April a public demonstration was held at Ballyclare in County Antrim. The first demonstration in England took place in May at Dormington Court Farm, Hereford, and was arranged by one of the first appointed Ferguson agents, Imperial Motors. The tractor was priced at £178, and although the hydraulic equipment was an integral part of the machine, it was quoted at £46 extra, making a total of £224.

To make full advantage of the tractor, a range of four Ferguson implements costing £26 each were offered. The two-furrow plough was designated the Type B, while the Type C was a spring-tine general purpose cultivator, later known simply as the tiller; the Type D was a ridger and the Type E a rowcrop cultivator.

The first Type A to be sold, tractor No.12, was bought by John Chambers's father in June 1936. The first production model, tractor No.1, was eventually sold in January 1937 to Thomas McGregor Greer for use on his Tullylagan Estate, near Cookstown in County Tyrone, in recognition of the landowner's financial support. McGregor Greer was also made a director of the sales company.

By September 1936, the tractor had been demonstrated as far afield as the Channel Islands and Harry Ferguson Ltd. claimed to have sold fifty machines. Production began slowly, but by the end of the year, 100 tractors had been built. Eventually, Park Works was turning out ten Ferguson-Browns a week.

The assembly shop for the tractors at Park Works was part of a five-storey building. The ground floor was taken up with a production line for gearboxes that were supplied to Scammel trucks. The main tractor assembly area was on the first floor. Here, the main components, such as the transmission and hydraulic units, were assembled at one of five separate stations, each with its own workbench. In front of these was the final assembly line that had the capacity for five tractors. The engines were run in for an hour by an electric motor drive, and then after an oil change they were tested for full power on a Heenan & Froude hydraulic brake dynamometer.

At Harry Ferguson's insistence, all the bolts used for the Ferguson equipment were made from high-tensile steel and were fitted with special cyanide-hardened nuts. Only two sizes of hexagon head were used, so one spanner would fit all the nuts and bolts. The bolt-making plant in the tractor assembly shop alone cost over £6,000.

The Ferguson-Brown implements were assembled on the second floor of the works building, while the third floor housed the pattern shop. On the top floor was the paint section, which

RIGHT:
A Ferguson-Brown prepared for sale in Herefordshire, possibly on the premises of Imperial Motors, one of the first Ferguson agents. Note the dog at the wheel!

ABOVE:
A Ferguson Type A tractor with the David Brown engine that was phased in from 1937. Note the early type of oil strainer with the brass handle that turned the filter against a brush. Fenders became available during 1938.

was quite automated for the time with a system of conveyors and turntables in the painting booths. All the tractors and implements received two coats of paint with a special rust-proofing primer. There was also a training department at the works for teaching farmers, operators and dealers the principles of operating Ferguson equipment.

Harry Ferguson aimed much of his advertising at the farms that were still using horses, which he maintained were 'picturesque but out of date'. He liked to point out that ploughing with draught animals was both laborious and costly. His favourite statement was 'One third of the land cultivated by horses is used for growing their fodder.'

The Ferguson-Brown was well received, and its superiority on steep ground meant that it sold well in hilly areas such as Scotland. It was even claimed that one farm eventually bought fourteen tractors. But in reality, overall sales were disappointing, as many farmers resisted buying a machine that needed its own special and costly implements. The Fordson was nearly £80 cheaper and could be used with any adapted horse-drawn or trailed equipment already in the yard.

By 1937, stocks of unsold tractors were mounting up. David Brown proposed a number of design changes to increase sales, but Harry Ferguson was vehemently protective of his 'baby' and refused to consider even the slightest change. As the sales company, Harry Ferguson Ltd., was running into financial difficulties, it was merged with David Brown Tractors Ltd. The new concern, known as Ferguson-Brown Ltd., was formed in June 1937. Harry Ferguson and David Brown were joint managing directors, but Ferguson held a minority shareholding and could see control of his tractor slipping away from him. Incidentally,

LEFT:
An example of the hop or orchard model Ferguson-Brown that was introduced in May 1938. It had a revised layout for the front axle that moved it back by 12 in. to give the tractor a smaller turning circle. Note the Standard Vanguard service van in the background as well as a Ferguson 3-ton trailer that must have been nearly new when the photograph was taken.

ABOVE:
John Chambers demonstrates the ability of the Ferguson Type A tractor to cope with steep hillsides to a group of farmers in Norway. The demonstration was held near Sola airport on 8 April 1938. Scandinavia was the only overseas territory in which the Ferguson-Brown achieved any success.

the general manager of Ferguson-Brown Ltd. was Walter Hill, who had been involved in importing Ferguson-Sherman ploughs for Muir-Hill during the 1920s. He was very experienced in tractor production and marketing, and had held various positions at Rushton, Roadless and Bristol Tractors between 1927 and 1935.

The engine fitted to the Type A was now of David Brown's own manufacture with castings made at Penistone. It seems that Coventry Climax had won a large contract from the Ministry for its Godiva fire pumps that became standard equipment for the British Army, Royal Air Force and Civil Defence units during the Second World War. The company was thus no longer interested in building small batches of engines for the Ferguson tractor and agreed to sell the casting machinery for the E Type to David Brown.

The David Brown engine was virtually identical to the Coventry Climax unit, but had a 3 ⅛ in. bore and a larger sump to increase oil capacity and aid lubrication when working on hillsides. The cylinder head was also different with the spark plugs evenly spaced, unlike the 'paired-plug' head used on the Coventry Climax engines. Other detail changes included an oil-bath air-cleaner in place of the large American dry air-filter supplied by Donaldson of Minnesota that had been a feature of the early tractors. An oil-strainer was also fitted with a self-cleaning mechanism, operated by a handle on top of the

RIGHT:
A late Ferguson-Brown tractor with the conventional oil filter that was one of the improvements initiated by David Brown. Harry Ferguson would not be pleased to see it working with a trailed cultivator.

LEFT:
The details of the main features of the Ferguson-Brown can be clearly identified on this excellent ⅙th scale model built by Jim Russell and Morris Fitchett. The model is of tractor No. 1081, built in about 1938.

filter-bowl that turned the filter against a brush.

There is some controversy over the stage at which the David Brown engines were actually introduced. A number of part changes were brought in at tractor No.256, and it seems likely that this was in readiness for the new power units to be phased in. There was obviously a transitional period when both makes of engines were used, and there is evidence that some David Brown blocks were fitted with Coventry Climax heads, and vice-versa, giving credence to the claim that no two Ferguson-Brown tractors were ever alike. It is believed that the gear ratios in the transmission were also revised at the same time as the change to David Brown engines. There was some talk of modifying the drive to the hydraulic pump to make it independent of the drive to the wheels, but it was decided that this would be too involved and costly.

By the end of 1937, the Ferguson Type A was offered with a choice of wheel equipment. Narrow 6 in. rear wheels for rowcrop work could be specified as an alternative to the standard 10 in. rear steels, or the tractor could be bought with 4½ x 19 in. front and 9 x 22 in. rear pneumatic tyres. Mudguards, or fenders, which had not been available for the first Ferguson-Brown tractors, were offered during 1938 and were described as 'wheel guards'.

A special hop and orchard model of the Ferguson Type A was introduced in May 1938. It was no more than a simple conversion, with the front axle moved back approximately 12 in. to give the tractor a smaller turning circle and make it more manoeuvrable for working in confined spaces. The company also marketed an industrial kit that consisted of larger rear wheels with 11¼ x 24 in. tyres and pressed-steel front wheels fitted with 6 x 19 in. equipment.

Sales picked up a little during 1938, and the company claimed to have sold 750 tractors that year, which was nearly half the total production run for the Ferguson-Brown. Export orders, however, were few and far between, and the North American market remained elusive, much to Harry Ferguson's disappointment.

The only really notable overseas sales were to Scandinavia. John Chambers was sent to Norway in April 1938 at the request of the local dealership to demonstrate the ability of the Type A tractor to cope with stony and hilly land. A successful demonstration elicited an order for twenty-nine tractors. One of these was later returned to the factory with a broken barrel housing, highlighting weaknesses in the strength of the aluminium alloy castings. Further problems with broken castings saw several tractors fitted with cast iron barrel housings as an interim measure while different specifications of alloys were investigated. Some hydraulic top-covers were also fabricated in cast iron.

David Brown was unhappy with the Type A's shortcomings and continually battled with Ferguson over the design. As Harry Ferguson refused to compromise, Brown pushed ahead with plans to build his own heavier and more powerful tractor. He also initiated several improvements to the Ferguson-Brown in an attempt to make it easier to sell, including a combined pulley and power take-off attachment, a conventional oil filter incorporating an oil pressure gauge, and a vaporiser to enable it to run on paraffin.

Although Harry Ferguson had resisted the use of paraffin as a tractor fuel, there was an increasing

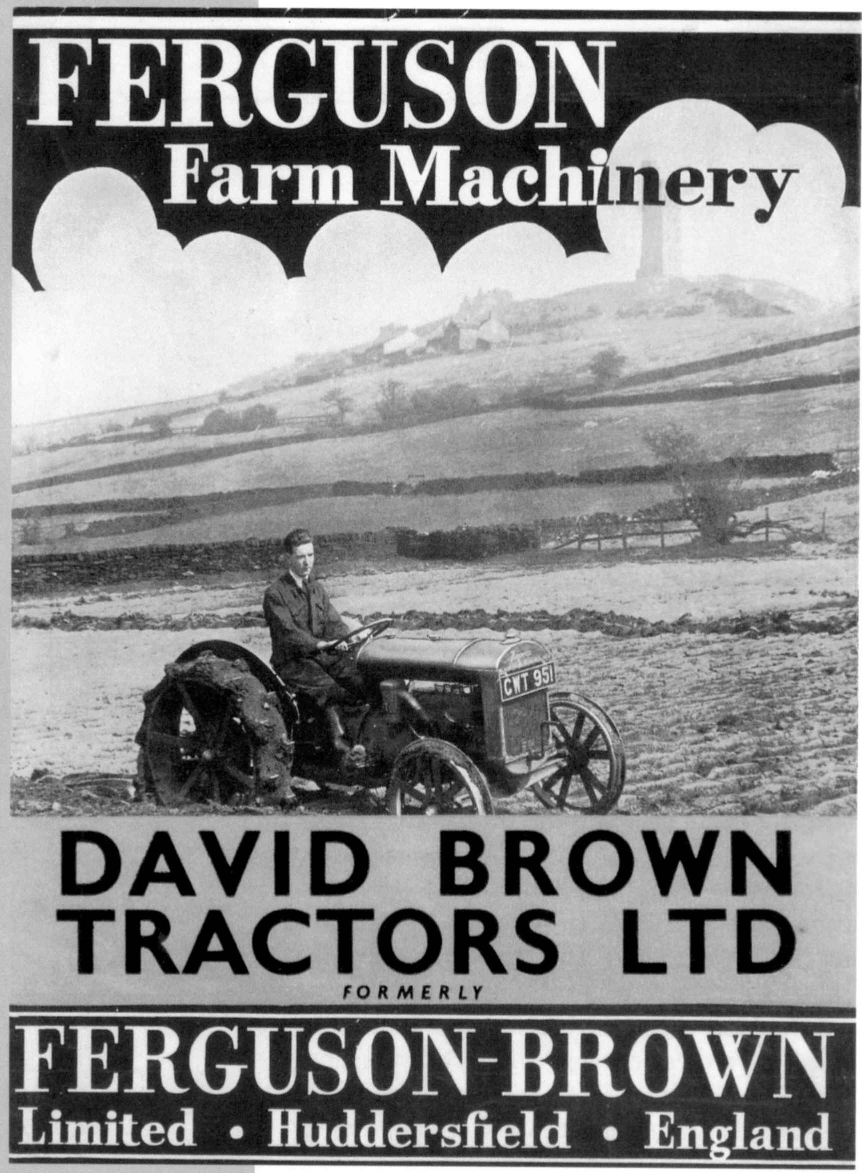

ABOVE:
A sales brochure for the Ferguson-Brown tractor. Note that it is overprinted with David Brown Tractors Ltd., the company that replaced Ferguson-Brown Ltd. from February 1939.

demand for a vaporiser attachment to allow the Ferguson-Brown to run on the cheaper vaporising oil. With the Second World War looming, petrol in Britain was becoming expensive and difficult to obtain. The vaporiser for the Type A was devised by Arthur Gladwell, and supplied by his company, Gladwell & Kell Ltd. of Ampton Street Works, Grays Inn Road, London. Gladwell was an inventor who had patented his first vaporiser in 1935. He claimed that he could make any engine run on any liquid that was inflammable, and during the Second World War, he was asked by the Ministry of Supply to adapt vehicles to run on creosote.

The vaporiser attachment, known as the Gladwell Kerosene Carburettor and Superheater, was offered as a kit that included a temperature gauge and cost £15. From November 1938, the vaporiser became approved Ferguson-Brown equipment. It was not very efficient, however, and the engine lacked power when running on paraffin or vaporising oil.

The same year, a single-furrow plough was added to the range of Ferguson-Brown equipment. A potato spinner of David Brown's design was demonstrated in September, but was not put into production until the company's new tractor appeared. Other implements, including a disc harrow, a powered cultivator and a cutterbar mower, were in the planning stage.

By the autumn of 1938, the partnership between Harry Ferguson and David Brown had reached breaking point and in October Ferguson went to the USA to see Henry Ford. It was a successful visit; he returned with an agreement for Ford to build a Ferguson tractor in America and was ready to terminate his association with David Brown.

Brown offered to buy out Ferguson's share in Ferguson-Brown Ltd., which was renamed David Brown Tractors Ltd. with effect from 1 February 1939. David Brown took over full responsibility for manufacturing, marketing and servicing the last of the Type A tractors. Production of the Ferguson-Brown continued for a short time until mid-1939. The very last few were assembled at David Brown's new tractor factory, the former United Thread Mills at Meltham, about three miles from Park Works. This was to be the home of the new David Brown tractor, the VAK1 model that was launched at the Royal Show, held at Windsor from 4 to 8 July 1939.

At the same show, the company exhibited three Ferguson tractors: a paraffin model on 10 in. steel wheels with a two-furrow plough, and two petrol tractors. One of the petrol tractors had pneumatic tyres and the new cutterbar mower, while the third machine was shown with narrow wheels and a rowcrop cultivator. It was a final last-ditch attempt to move the remaining stocks of the old model Ferguson-Brown.

Over 1,350 Ferguson Type A tractors were built. There are no exact figures, but the latest Ferguson-Brown in existence bears the commission No.1354, and was probably one of the last to be made. It was not a great number and the tractor certainly did not make the impact expected or deserved. It could be argued that there were too many limitations to the Type A; it was expensive to buy and equip, it lacked power when running on paraffin and suffered from too many expensive failures caused by stress fractures to the aluminium castings.

But, for all this, it was still a landmark machine. It was the first tractor to incorporate a three-point linkage, hydraulic lift and an effective system of depth control and as such it was an important milestone in the development of mechanised farming. It also marked the beginning of two famous tractor lines, Ferguson and David Brown. Because of this historical significance and the relative scarcity of Type As, Ferguson-Brown tractors are highly sought after today and command high prices, while the Ferguson 'Black Tractor' is rightly preserved in the Science Museum in London.

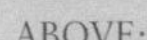

ABOVE:
A Ferguson-Brown tractor on demonstration with a prototype David Brown potato spinner in September 1938.

LEFT:
A Ferguson-Brown tractor. The Type A remained in production until mid-1939 and just over 1,350 were built.

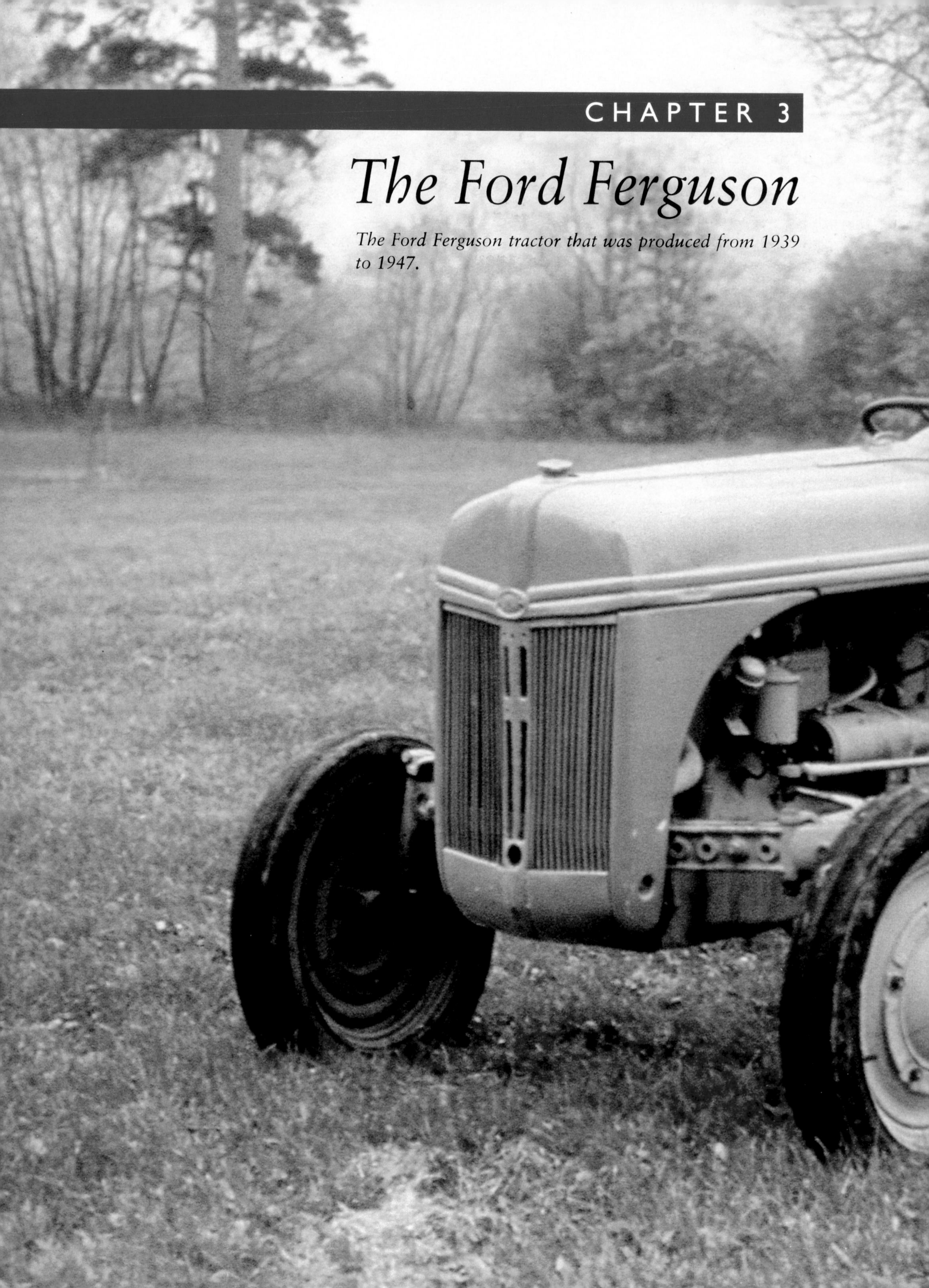

CHAPTER 3

The Ford Ferguson

The Ford Ferguson tractor that was produced from 1939 to 1947.

ABOVE: *One of two Ferguson-Brown tractors that were shipped to the USA in October 1938 for the purpose of demonstrating the Ferguson System to Henry Ford. The demonstrations were arranged with the assistance of the Sherman brothers and the first was held on the Ford family farm at Dearborn. In typical Ferguson fashion, a young boy is used to show how easy the tractor is to drive.*

Harry Ferguson's meeting with Henry Ford in October 1938 could hardly have come at a better time as both men were ready to commit themselves to a collaboration that might not have happened at any other period. Ferguson, resigned to the fact that his agreement with David Brown was turning sour, was looking for a new partner, and Henry Ford was ready to return to tractor manufacture in America.

The Ford Motor Company had not built any tractors in the USA since the Model F Fordson went out of production in 1928. The Model N Fordson was still being made in Britain at the Dagenham plant, but it had made little impact on the North American market and very few were imported. Henry Ford had been brought up as a farm boy, tractors were his first love, and he dearly wanted to see one bearing the Ford name back in production in the USA.

Ford began experimenting with tractors again in 1931. Between 1936 and 1938, several prototypes were built and tested. They can only be described as weird and wonderful and were not very successful. They were based on automotive components cannibalised from Ford cars and trucks. One was the archetypal 'flying bedstead' – a three-wheel tractor with one-wheel drive.

Eber Sherman of the former Ferguson-Sherman company was shown the Ford prototypes; he had seen the Ferguson Type A working in Britain and knew which machine offered the most potential. He told Ford of Ferguson's new tractor and engineered a meeting between the two men.

Ferguson sailed to New York in October 1938, taking his old friend, John Williams, to act as his assistant. The Sherman brothers arranged the trip, and Ford agreed to pay for it. Two Type A tractors, Nos. 661 and 662, and a selection of implements were crated and shipped to America for demonstration purposes.

LEFT:
The meeting between Harry Ferguson and Henry Ford that resulted in the famous 'handshake agreement' whereby Ford agreed to manufacture a tractor incorporating Ferguson's hydraulic system.

LEFT:
The meeting in October 1938 between Harry Ferguson and Henry Ford was held at Ford's Fair Lane residence. Harry's daughter, Betty, and Eber Sherman were on hand to witness the event while a photographer recorded the proceedings. Note the Ferguson plough, ridger and cultivators set out for the demonstration.

ABOVE:
An illustrious group gathered with what appears to be a pre-production model of the Ford 9N tractor, not yet bearing any identification badges. Left to right are John Chambers, Willie Sands, John Williams, Harry Ferguson, Henry Ford and Eber Sherman.

The meeting took place at Henry Ford's home, Fair Lane, near Dearborn in Michigan, and resulted in the famous 'handshake agreement'. Ford was suitably impressed by the Type A's performance and agreed to manufacture a tractor incorporating the Ferguson hydraulic system, and allow Ferguson himself to handle the sales and distribution. The agreement was no more than a verbal understanding between the two men, sealed with a handshake. It was never witnessed, no formal or written contracts were ever drawn up, and the exact terms would always remain a cause for speculation. It was rumoured that Ford agreed to buy the two Ferguson-Brown tractors for his farms as part of the deal to help Ferguson with his immediate expenses.

Ferguson returned to Britain to terminate his partnership with David Brown, elated that his system was about to go into mass production with the prestige and power of the mighty Ford organisation behind it, and that he was collaborating with one of the world's greatest industrialists. Sadly, like many of Harry Ferguson's associations, it was an ill-fated partnership that would end in acrimony in less than nine years.

While Ferguson was away sorting out his affairs, a team of engineers under Charles Sorensen began working on the new tractor almost immediately. One of the Type A tractors that had been brought over for the demonstration was stripped out in a corner of the Rouge River plant, Ford's giant factory complex, located on a 1,000 acre site south-east of Dearborn. Here, the Ford team began to examine, modify and evolve the design for mass production.

Two lightweight experimental tractors were built to evaluate the hydraulic system. Both were basic machines with two-speed gearboxes. One was powered by a horizontally opposed, two-

LEFT:
This photograph of a pre-production or demonstration model Ford 9N is designed to show that the tractor is light enough for a child to handle it.

cylinder, air-cooled engine, while the other used a 24 hp four-cylinder unit that was produced in Dagenham for the English Ford Model B truck.

The proposed design for the production machine was finalised after Ferguson returned to the United States in January 1939, bringing with him Willie Sands, Archie Greer and John Chambers to head his engineering group. Naturally, there was some conflict between Harry Ferguson and Ford's designers; Ferguson was dismayed to find that Ford's team, led by a young engineer called Harold Brock, had altered much and even changed the design of his plough, but Sorensen was not easily bullied and stood his ground. There was no love lost between the two men; Sorensen was very protective of Ford and probably regarded Ferguson as a threat to his position. Later, in his autobiography, he was to remark of Ferguson rather unfairly that, 'Henry Ford had raised him from a poor man looking for a job to a successful businessman with the Ford Motor Company.' Not exactly true.

In keeping with Ford's policy and to minimise production costs, a number of automotive components were incorporated into the new tractor. The engine was a 120 cu in. (1,966cc) four-cylinder side-valve unit developing 23 bhp. It was based on a Ford V8 engine, as used in the company's prestigious Mercury car range, that was simply sliced in half. The tractor had full electrics, with coil ignition, a battery, generator and self-starter. Its transmission was a three-speed unit mated to a spiral bevel-gear final drive with a four-pinion differential based on components taken from a 30 cwt Ford truck axle.

BELOW:
The Ford 9N's four-cylinder petrol engine was a side-valve unit that developed 23 bhp. Its design was based on half of the Mercury V8. Note the Marvel-Schebler carburettor.

A new design of extendable front axle set the precedent for all future Ferguson tractors. It was swept back and supported by two radius arms and had twin drag-links to allow track width to be altered quickly and easily. The rear wheels were also adjustable and formed from pressed steel discs for strength and lightness. Pneumatic tyres were standard equipment and independent drum brakes completed the specification.

The hydraulic system was refined and improved. The four-cylinder piston-pump took its drive from the gearbox countershaft that also drove the power take-off shaft through a dog-clutch. This meant that the hydraulics would work independently of the tractor's movement and would still operate when it was stationary.

The overall dimensions of the tractor were evidently determined, at Ferguson's suggestion, by the need to fit fourteen machines into an American railroad boxcar to reduce shipping costs. Edsel Ford suggested the shape of the tractor, and the sheet metal was designed by Ford's automotive stylists.

BELOW: *Harry Ferguson and Henry Ford with the Ford 9N tractor possibly at the public launch and demonstration held on the Ford estates at Dearborn in June 1939.*

The first prototype was ready by 1 April 1939, and was demonstrated in Clara Ford's vegetable garden. Other prototypes followed, and on 27 April, Henry Ford announced to the press his intention to build a new tractor incorporating Harry Ferguson's system. Once the design was finalised, it was put into production in the B Building at the Rouge River plant in June. The new tractor was given the model number 9N; the 9 signifying the year of introduction, 1939, and the letter N being Ford's tractor designation. The front of the hood carried the oval Ford badge, underneath which was fixed a plate that bore the legend 'Ferguson System' to satisfy Harry Ferguson.

The tractor was unveiled to the dealers at the Dearborn Inn on 12 June 1939, followed by a public launch and demonstration on Henry Ford's estates at Dearborn on 29 June in front of over 400 journalists, agricultural officials, businessmen and foreign government representatives. Introduced with the tractor were four different models of ploughs, a general-purpose cultivator and an inter-row cultivator supplied by the Ferguson-Sherman Manufacturing Corporation. This concern, based in Dearborn and established on 16 May 1939 with a loan of $50,000 from Henry Ford,

was set up by Harry Ferguson in conjunction with the Sherman brothers to handle the sales and distribution of the tractor and to manufacture the implements.

The arrangement continued until 24 April 1942, when Harry Ferguson fell out with the Sherman brothers over marketing problems and reorganised the company as Harry Ferguson Incorporated, appointing Roger Kyes as manager. Kyes, who had formerly worked for the Empire Plow Company before joining the Ferguson-Sherman concern in 1939, was an able businessman who proved adept at persuading other manufacturers to tailor their implements to suit the Ferguson tractor.

Sometimes known as the Ford Ferguson, the 9N was priced at $585 and met with immediate acclaim. Its automotive styling and lightweight design made it an attractive machine that was pleasurable to drive. It had everything the customer needed in a compact package with additional modern features such as an exhaust muffler and a safety starter switch. Even though some American farmers were sceptical of the advantages of the Ferguson System, over 10,000 9Ns were sold in the first year.

Problems with the steel presses and patterns meant that the first 700 or so tractors were fitted with aluminium hoods. Changes were also made to the hydraulic system to rectify small fluctuations of the draft control caused by the flexing of the rubber tyres. Production began slowly, but by 1941 Ford was making more than 40,000 units a year. Only a few were exported

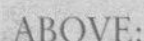

ABOVE:
The Ford 9N tractor, better known as the Ford Ferguson. Note the quadrant lever to control the hydraulic system.

LEFT:
An early Ford 9N model. Around 700 of the first Ford Ferguson tractors were fitted with aluminium hoods as the steel presses and patterns had not been ready in time for the start of production.

RIGHT: *A 1939 Ford 9N harvesting with an Allis-Chalmers All-Crop combine.*

and these were sold to Alaska, Canada, Cuba, Mexico and the Hawaiian islands.

Henry Ford, like Ferguson, was a visionary and saw in the 9N an opportunity to once again revolutionise mechanised farming in America as he had done with the Model F Fordson twenty years earlier. The early advertising literature bore the imprints of both men, and claimed that the tractor would cut costs, make farming more prosperous, stop the drift of youth from the land and 'lay the foundation for a greater National Security'. Bold claims for a little grey tractor.

Harry Ferguson also joined with Henry and Edsel Ford to set up, promote and sponsor the National Farm Youth Foundation. This movement was established to educate and train young Americans in farm management and machinery operation and maintenance. It goes without saying that all the courses were based on the use of Ford Ferguson tractors and equipment. From 1941, the foundation membership was opened up to women, and after the USA entered the Second World War, they became the American equivalent of Britain's Land Army girls when the conflict led to labour shortages on farms.

A vaporising oil version of the new Ford tractor, designated the 9NAN, was introduced

RIGHT: *A Ford Ferguson at work. Over 10,000 9N models were sold during the tractor's first year of production, and by 1941 Ford was making over 40,000 a year.*

ABOVE:
A Ford 9NAN model designed for the British market. Fitted with a vaporiser for running on vaporising oil, it was launched in October 1939.

for the British market. Ferguson had evidently persuaded Henry Ford to agree to his tractor going into production at Dagenham, but he was met by a wall of opposition from the board of the British Ford Motor Company who were understandably opposed to the idea. Dagenham was already committed to a massive wartime production run for the Model N Fordson, and the time and expense involved in re-tooling for a new tractor at this critical time made the move impossible.

Not wanting to be denied his chance to re-launch a Ferguson tractor in Britain, Ferguson staged several demonstrations of the 9N across the UK from October 1939, culminating in an official press introduction of the tractor and its implements at St. Stephen's, near Bedfont in Middlesex, in May 1940. The object of the exercise was to undermine the opposition of the British board under Lord Perry to production in the UK. However, the Ford Motor Company in Britain was answerable to its own shareholders and, enjoying a certain degree of autonomy from the American parent organisation, was able to ignore Ferguson's exhortations.

Some 10,000 of the American Ford Fergusons were eventually sent to Britain under the Lend-Lease agreement during the Second World War. The tractors were shipped into the UK in knocked-down form packed in crates; just over half of the total consignment was re-assembled at Dagenham while Harry Ferguson Motors in Belfast handled the remainder.

The tractors that went to Northern Ireland arrived in batches of 500 every month to six weeks. Because of the pressure of work, the staff at the May Street premises worked overtime and the Ford Fergusons were put together in the evenings. Two men could unpack and assemble two tractors in two hours. The Donegall Square East showrooms were cleared of cars and the completed tractors were stored there until the War Agricultural Committee assigned them to farms. Most of the Ford Fergusons imported were the later 2N models that came in towards

ABOVE:
Girls of the National Farm Youth Foundation ploughing with Ford Ferguson tractors. This movement was jointly sponsored by Harry Ferguson, Henry and Edsel Ford.

RIGHT:
The Ford 2N utility model was fitted with steel wheels and magneto ignition, and had to be started by hand. It was introduced in 1942 following material shortages in the USA.

the end of the conflict. In 1942, the standard agricultural 9NAN cost £310 in Britain, while the 2N utility model was priced at £282.

The 2N model Ford-Ferguson was originally introduced as a utility model in 1942, hence its designation. Material restrictions and steel shortages were affecting American industries following the entry of the USA into the war in December 1941. This meant that Ford had to cut back on the production of the 9N and re-launch it as the 2N without electrics and fitted with steel wheels and magneto ignition. Within a year, the restrictions were eased and the 2N reappeared on rubber tyres with its full

LEFT:
After the material restrictions and steel shortages eased, the Ford 2N reappeared with its rubber tyres, electric start and coil ignition restored. The only way to tell it apart from the 9N model it replaced was by the long slots cut into the centre bar of the front grille.

electrical system including a starter and coil ignition restored. Some minor design improvements were made, but it was little different to the 9N it replaced. Outwardly, the two models could only be told apart by the long slots cut into the centre bar of the front grille fitted to the 2N.

Industrial versions of the 9N/2N design included the Ford Moto-Tug, a heavy-duty towing tractor for haulage, moving aircraft or handling shipping containers in dockyards. Equipped with steel plate pusher bumpers front and rear, the Moto-Tug was available in two versions; the B-NO-25 had a rated drawbar pull of 2,500 lb, while 4,000 lb loads were within the capacity of the heavier B-NO-40 model that was fitted with dual wheels. The Moto-Tugs were extensively used by the United States Air Force, and at least two saw service in Britain during the Second World War.

Harry Ferguson never gave up hope of seeing the Ford Ferguson going into production in Britain, and held out for a change of heart by the British Ford Motor Company once the war was over. Willie Sands had been keen to return to Belfast, and was followed by Archie Greer as soon as the design for the 9N had been finalised, leaving John Chambers to act as chief engineer in the USA. They both continued working on prototype implement components and modifications in Ireland. The aim was to produce a range of British implements for the Ford Ferguson, but wartime material shortages were making this a difficult task.

Harry Ferguson Motors Ltd. in Belfast was regarded as a prestigious company and apprenticeships with the firm were highly sought after, even though it charged £100 for the privilege, which was the price of a new Ford car. In 1938, there were 500 applicants for just five vacancies advertised in May. In the event, only two apprentices were taken on, and one of these, Alex Patterson, was seconded to assist Sands and Greer, the three of them working in cramped conditions in a small attic room over the Donegall Square premises with a single drawing board.

Because there were no cars to sell during the war, the company moved into armaments manufacture, making Bofors anti-aircraft guns and templates from 1941 in a small factory at Moira, a village near Lurgan, about thirty miles from Belfast. It was here that some of Sands' and Greer's implement designs were translated

RIGHT:
A Ford 2N tractor with a front blade fitted with side plates for snow clearance.

into metal, and enough steel was procured to make a small number of potato spinners and a steerage hoe, known as a sugar beet cultivator, that were sold bearing 'Harry Ferguson (Motors) Ltd.' badges.

Some of these Irish implements for the Ford Ferguson were shown by the Ford Motor Company at a demonstration of rowcrop equipment held on 19 and 20 April 1944 by the Highland and Agricultural Society of Scotland at a farm in Castleton, near Eassie in Angus. A tiller, spring-tine cultivator and tandem disc harrow, virtually identical to those later introduced for the TE-20, were shown along with the sugar beet cultivator. A cutterbar mower of American parentage completed the range.

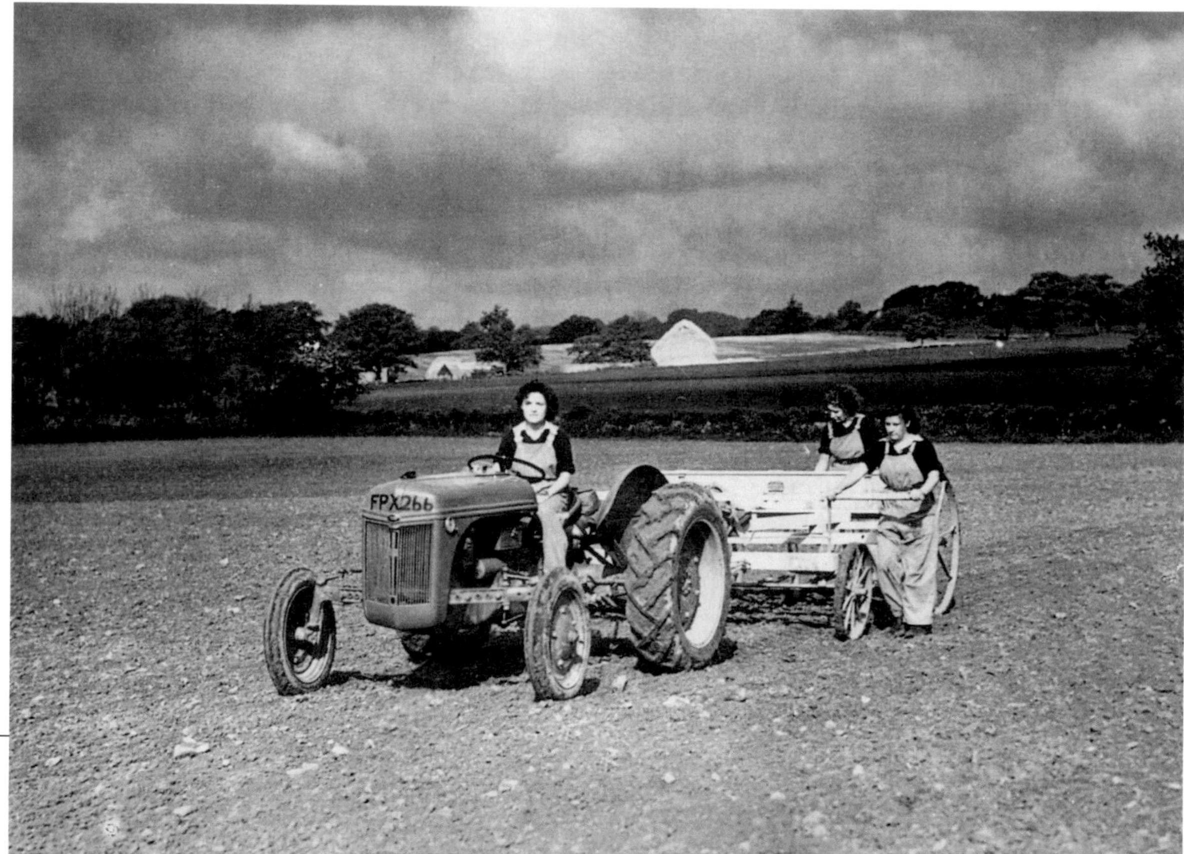

RIGHT:
Land Army girls using a 1942 Ford 2N for autumn drilling in Britain during the Second World War.

Sadly, within a little over a year, the Ford Motor Company seemed to have lost all interest in promoting Ferguson implements. Plans for a new model Fordson, the E27N Major, were underway, and at a press luncheon held in London on 14 June 1945, Lord Perry announced that Ford in Britain was to begin manufacturing implements for Fordson tractors. In the same speech, he finally dashed Ferguson's hopes by categorically stating, 'The company are not going to make the "Ford with Ferguson System" tractor.'

Harry Ferguson was deflated but not undeterred and forged ahead with his own plans to build Ferguson equipment in Britain. A few days after Perry's announcements, Ferguson issued a statement from America saying that he was coming to England to set up 'the largest manufacturing industry in Europe' and planned 'the production of 1,000 tractors and 10,000 implements a day'. The press release also stated that the cost of the tractors was estimated at £100 each. It was an over-ambitious and unrealistic statement, but Ferguson was typically piqued; he was going to take the Ford Motor Company on at their own game, and planned to out-produce and undercut them.

The final variation on the Ford Ferguson design was the 4P, an experimental and more powerful version of the tractor, designed to meet the demand for a larger machine. Two separate teams from the Ford Motor Company and Harry Ferguson Incorporated worked on the design. It is a standing joke that many tractors are designed by a committee, but this must be the first time a tractor was conceived by two committees from two separate companies!

The first prototype 4P appeared in 1944 and was powered by a four-cylinder, side-valve Continental petrol engine. It had more ground clearance than the 9N/2N and was fitted with 9 x 36 in. tyres. Another four were built over the next couple of years; one was sent to Ferguson's new British company in Coventry in 1945, while another went to Dagenham in 1946 for evaluation. The tractor seen at Dagenham was probably the last example. It was fitted with an overhead valve engine that was unlike any proprietary make. There is no record of the 4P design going beyond the experimental stage on either side of the Atlantic, although ironically both the Ford

Until then . . . A day will come when with a handshake the farmer will pay his tribute to the work the Land Girls did to keep things going. Until then, the vast mechanised army of farm workers is winning the battle against the blockade. And Ford and Fordson dealers are playing a big part by helping farmers to keep their tractors at work. The Ford repair and maintenance service cuts out delays and hold-ups on the farm—and saves fuel for the nation.
WE WON'T WASTE IT, SAILOR
Farm by Ford or Fordson
FORD MOTOR COMPANY LIMITED, DAGENHAM ESSEX. LONDON SHOWROOMS: 88, REGENT STREET, W.1

LEFT: *A 1944 British advertisement for the Ford 2N. Over 10,000 Ford Ferguson tractors were exported to the UK under the Lend-Lease agreement during the Second World War.*

BELOW: *A rather battered wartime Ford 2N tractor that has been converted to diesel power and fitted with a Perkins engine is seen mowing hay in the Peterborough area.*

and Ferguson companies separately evaluated the tractor for possible production in England at a later date.

The American Ford Ferguson partnership eventually became a casualty of political upheaval within the Ford organisation. Henry Ford's son, Edsel, who had taken over from his father as president of the Ford Motor Company in 1918, died of cancer in May 1943, aged only 49 years. Henry had to reassume the mantle of head of the company. Now 80 years old, failing in health and having suffered a series of strokes, he no longer had the strength or full presence of mind to run the giant organisation. Following a series of power struggles that saw the resignation of Charles Sorensen as vice-president in May 1944, the company started to flounder and accumulate serious losses. The crisis was only resolved after Henry Ford was persuaded to hand over to his grandson and Edsel's eldest son, Henry ll, with effect from 20 September 1945.

Henry Ford ll was only 28 years old when he took over the reins of the Ford Motor Company, but he was a strong character and quite capable of instituting any necessary

ABOVE: *An industrial version of the Ford Ferguson tractor. Known as the Ford Moto-Tug, it was a heavy-duty towing tractor for factories, docks or airfields. The tractor shown was used in Britain by the USAF for moving aircraft during the Second World War.*

RIGHT: *The experimental Ford 4P tractor, a larger and more powerful version of the Ford Ferguson. Five prototypes were built between 1944 and 1946.*

LEFT:
A version of the Ford 4P was sent to Dagenham for evaluation in 1946. This view of the driving position appears to show a column-mounted gear-shift. Note the rear-mounted belt pulley attachment.

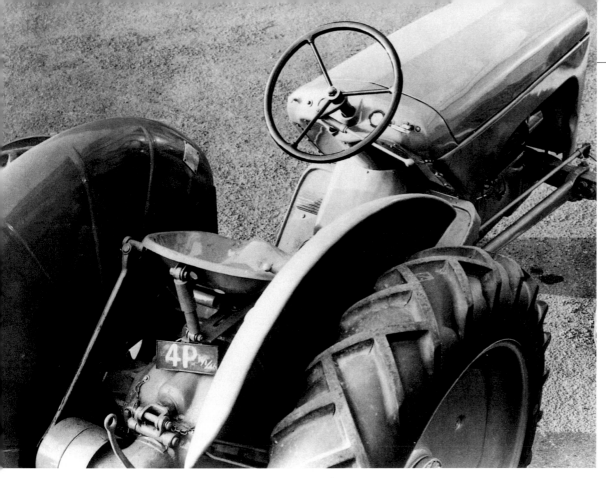

changes to revive the ailing business. One of his first targets was the tractor operation, which had made estimated losses of around $20 million.

Wartime price controls had drastically reduced the Ford Motor Company's returns from manufacturing the tractor. Ferguson's cut through his distribution arrangement was not so badly affected and on paper his company was making close to $4 million a year. It looked as if he had made more out of the agreement than Ford and rightly or wrongly, he was blamed for much of the losses. There was considerable resentment on both sides as yet another of Ferguson's partnerships began to fall apart.

In June 1945, Henry Ford II moved production of the Ford Ferguson tractor to the Highland Park plant outside Detroit, and then in November 1946 he terminated the relationship with Ferguson and announced that the tractors would be distributed exclusively through the Dearborn Motors Corporation, a new Ford subsidiary company. Production of the 2N would continue until the following summer when it would be replaced by a new Ford model.

The final Ford Ferguson came off the line in July 1947. A total of 306,256 9N and 2N models had been made over eight years and, during the same period, the Ferguson company

BELOW:
A Ferguson sugar beet cultivator with a Ford 2N at a demonstration in Scotland in April 1944. It was one of a small number of implements made in Ireland by Harry Ferguson (Motors) Ltd. at a factory in Moira.

RIGHT: *A Ford 2N at the end of the Highland Park assembly line in August 1946. Ford Ferguson production had been transferred to the Highland Park plant in June 1945.*

had sold 943,957 implements. The total sales figures for the tractors, implements, accessories and parts grossed Harry Ferguson Incorporated over $312 million.

Although the Ford Ferguson did not make a lot of money for the Ford Motor Company, it was an unrivalled success in terms of the impact it had on American agriculture. It brought a first taste of mechanised farming to many of the smaller family farms in the USA and inspired many of the designs that followed. The affection for this diminutive little tractor is still undiminished in the USA after over fifty years.

Ford's replacement for the 2N was outwardly a very similar machine and was known as the 8N. The most obvious difference was a change in colour from the battleship grey of the Ford Fergusons to a new vermilion red and light grey livery. Sales literature for the 8N boasted twenty-two new features, but many of these were minor modifications. The most significant design improvement was the provision of a four-speed gearbox. At the heart of the tractor was what was controversially referred to as the

RIGHT: *One of the last FordFerguson tractors – a 2NAN model that was not registered in Britain until 1948. Production of the 2N had ended in July 1947 after over three million Ford Fergusons had been built.*

LEFT:
The Ford 8N – the tractor that replaced the 2N in 1947 and sparked off a lengthy court battle between Harry Ferguson and the Ford Motor Company.

Ford Hydraulic System. This was based on Ferguson's system with a few alterations and the inclusion of position control. Although originally planned for release in 1948, hence the 8N designation, the tractor was actually rushed into production in July 1947 after the 2N was dropped following the break up of the relationship between Ford and Ferguson.

To say that Harry Ferguson was not a happy man due to this turn of events was an understatement; his American distribution organisation of thirty-three distributors and 2,876 dealers faced closure with no tractors to sell, and the new Ford tractor would evidently still use his hydraulic system without any concession to paying him royalties on his patents. His plans to build a new Ferguson tractor in Coventry in England were well underway, but he was not prepared to let Ford off lightly. In January 1948, he filed a lawsuit against the Ford Motor Company for the alleged ruination of his business and patent infringement. He claimed over $251million in damages plus costs.

The ensuing legal battle was prolonged and vicious. It lasted until April 1952 when Ford agreed to pay an out of court settlement of $9.25 million, and to cease using Ferguson patents. Ironically, to get around the patent issue, all Ford had to do to the 8N's successor, the NAA tractor that was announced in January 1953, was to move the hydraulic control valve from the suction to the delivery side of the pump. Both Ford and Ferguson claimed the outcome of the court case was a victory, but in truth Harry Ferguson was privately disappointed with the settlement. His one consolation was the amount of publicity it had generated for his new British tractors.

CHAPTER 4

The TE-20 Family

The Ferguson TE-20 was produced in several versions at Banner Lane from 1946 to 1956. Over half a million were built.

Harry Ferguson began looking for a British manufacturer for his tractors as early as 1943 once it became evident that the Ford Motor Company was unlikely to put the 2N into production at Dagenham. Initially, the search for a suitable company was left in the hands of Trevor Knox. After Lord Perry's announcement in June 1945, leaving no doubt that Ford had no intention of building a Ferguson tractor in the UK, Ferguson crossed the Atlantic to speed up the quest to find new British production facilities.

Knox had already made tentative contact with Sir John Black, the deputy chairman and managing director of the Standard Motor Company based in Coventry. At the age of 52, Black was already a very respected figure within the motor industry. He had built his first petrol engine at the age of nineteen, just before the outbreak of the First World War, during which he served in Gallipoli and France. Demobilised with the rank of captain in 1919, he joined the early car manufacturer, ABC, before transferring to the Hillman Motor Car Company in Coventry later that same year. Following the death of William Hillman in 1921, he gained a controlling interest in Hillman in partnership with his brother-in-law, Spencer Wilks, before selling out to the Rootes Group in 1929. Wilks then moved to Rover at Solihull, becoming managing director, while Black stayed in Coventry to take up a position with the ailing Standard Motor Company at Canley.

Standard was founded in 1903, the year it built its first car, the 'Motor Victoria' with a 6 hp single-cylinder engine, in premises at Much Park Street, Coventry. By the end of the 1920s, the company had expanded into the Canley site and was building a range of models, but was struggling financially and was in debt to the bank and creditors to the tune of nearly £500,000. John Black came in with an

The Banner Lane plant in Coventry as it was in 1946 at the beginning of Ferguson production. Built in 1938, it was originally used as a shadow factory for manufacturing aircraft engines, and still bore its wartime camouflage markings when the photograph was taken.

ABOVE:
Harry Ferguson and Sir John Black with one of the first TE-20 tractors. The manufacturing agreement between Ferguson and the Standard Motor Company was signed on 20 August 1946.

innovative reconstruction plan involving modern initiatives such as sourcing components from outside suppliers and introducing productivity bonuses for the workforce. Within a year, he had turned the company around; it was out of the red and was showing a £40,000 profit. He was appointed managing director in 1934, and introduced many new and successful models through the 1930s, including the famous 'Flying Standards' launched in 1936.

During the Second World War, the Standard company became involved in aircraft production, operating two shadow factories leased from the government. The larger of these, built in 1938 at a cost of £1.7 million on an 80 acre site at nearby Banner Lane, had a floor area of over 1 million sq ft and was used for the manufacture over 20,000 Bristol aero-engines. The other shadow plant at Ansty produced around 1,000 Mosquito fighter-bombers. In 1941, Black became chairman of the joint Aero-Engine Shadow Committee of the Ministry of Aircraft Production and he was knighted for his work two years later.

After the war, with Sir John Black's vision and forthright leadership, the company moved forward, championing modern production methods and working practices, and was the first British car manufacturer to introduce a five-day week. Like other manufacturers of the immediate post-war period, Standard was still saddled with pre-war designs, notably the Eight, Twelve and Fourteen models. Black decided to simplify production and cut costs by instigating a one-model policy and concentrating on a new design of car that could compete on world markets.

The planning of the new car, which became the Standard Vanguard, began in 1945. It was to be a medium-powered model with streamlined bodywork and was designed to appeal equally to the home, American and other

ABOVE: *The headquarters of Harry Ferguson Ltd. on Fletchampstead Highway. The premises, originally Standard's old Fletchamstead North Works, housed Ferguson's administrative offices and engineering department. Harry Ferguson's Rolls-Royce can just be seen behind the bus in front of the main door.*

export markets. Like the Ferguson tractor, it was one model for many purposes.

Black was more than a little interested in producing the Ferguson tractor, as its concept matched his engineering ideology. His company had the capacity to build the tractor in the former shadow factory that now lay empty at Banner Lane, and he could foresee that there could be a possible interchange of components with the Vanguard car to keep costs down. The problem was funding the operation and overcoming the steel shortages.

For men with Harry Ferguson and Sir John Black's energy and tenacity, no problem is insurmountable; demonstrations of the Ferguson System were laid on, and Sir Stafford Cripps, the president of the Board of Trade, was badgered until the government gave in and sanctioned the necessary loan and materials for tractor production to begin. Cripps was so won over by Ferguson's enthusiasm that he actually advised Standard to build the tractor.

The final hurdle was getting Ferguson and Black to agree the production details with each other. An alliance of two men of such strong character and similar conviction was bound to lead to minor arguments and disagreements. Press announcements to the effect that Standard was to make the Ferguson tractor appeared in November 1945, but it was the following summer before all the differences between the two parties were finally resolved, mainly due to the intervention of Ferguson's American manager, Roger Kyes. It was also an expensive time for Standard; in December the company bought the rights to another Coventry car manufacturer, aquiring the Triumph marque for £750,000, and during the next two years it spent over £3 million on re-tooling Banner Lane for Ferguson production and adapting Canley for the Vanguard.

The agreement between Ferguson and the Standard Motor Company was signed on 20 August 1946. Standard was to manufacture the Ferguson tractor for ten years, while Harry Ferguson set up an independent company to undertake the marketing, design, research and development of the tractor and implements. The new company was known as Harry Ferguson Ltd. It joined its American counterpart, Harry Ferguson Incorporated, under the umbrella of Ferguson Holdings Ltd. Charles Vincent, who had been on the board of the American concern, was appointed managing director of Harry Ferguson Ltd., while Roger Kyes became managing director of the holding company.

Harry Ferguson Ltd. operated out of premises leased from the Standard Motor Company on Fletchampstead Highway. These premises adjoined the Canley site and were originally the Fletchampstead North Works. This was a shadow factory built over Standard's old works golf course, just off Tile Hill Lane, for the manufacture of Oxford training aircraft during the war.

The Fletchamstead site, affectionately known as 'Fletch' by its staff, not only housed the administration, sales and design offices, but also had workshops for the engineering department and a fully equipped test station with electric and air-brake dynamometers. To oversee the development of the new British product line, Harry Ferguson moved back to Britain and bought a large mansion called Abbotswood, set in 600 acres of pasture land near Stow-on-the-Wold. Situated in the picturesque Cotswolds, it was also in easy commuting distance of his Fletchampstead headquarters.

Work on designing the new Ferguson tractor and implements had begun in 1945. Willie Sands and Archie Greer moved to Coventry to take charge of the designs, bringing Alex Patterson with them. John Chambers had submitted some prototype models and plans from the American team, including a 4P tractor sent for evaluation, but all of these were scrapped in favour of starting from scratch with a completely new British design. This decision was made partly because Harry Ferguson felt that he had not had enough involvement in the American proposals.

The new tractor was based loosely on the Ford-Ferguson design, but much was changed. Sands and Greer did most of the design work;

LEFT: *One of the first Ferguson TE-20 tractors working at Hawkes Moor Farm at Berkeswell, behind the Banner Lane works. The farm belonged to the Standard Motor Company and was tenanted by the Williams family whose son, Les, is seen at the wheel. Launched on 18 September 1946, the TE-20 was powered by a Continental petrol engine and developed 23 bhp.*

they understood Harry Ferguson better than most and shared his ideology. He insisted that every ounce of weight saved meant less fuel used and more profit for the farmer. It was a view that both Sands and Greer could sympathise with, well aware of the impoverished state of farming back in Northern Ireland.

All the drawings for the new tractor were drawn twice: once by Ferguson's draughtsmen and then to suit Standard's filing system. Standard's engineers also had some input on the design as it had to suit their production methods, and even Sir John Black took a close interest in the project.

BELOW: *A cutaway view of the Ferguson TE-20 tractor.*

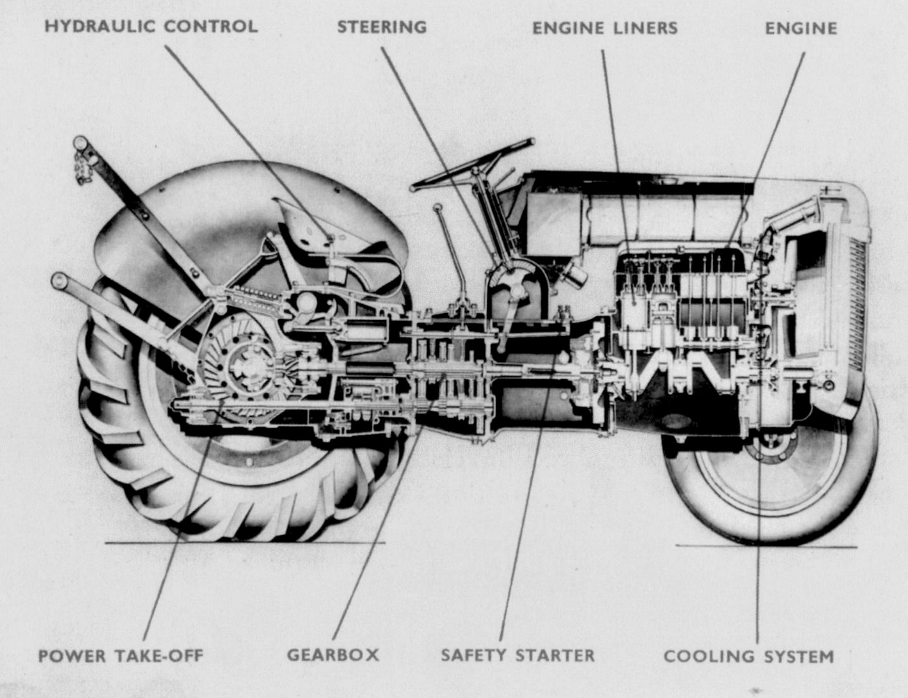

One of the limiting factors in the development of the tractor was the size of the rear axle. It provided the mounting points for the three-point linkage and could not be changed or the older Ferguson equipment would not fit and the complete implement line would have to be redesigned. The problem was that its design was carried over from the light-truck axle used in the Ford Ferguson, thus restricting engine power to around 20 hp.

Standard began to work on a suitable engine, but as an interim measure power units were sourced from the USA. They were supplied by the Continental Motors Corporation of Muskegon, Michigan, which claimed to be 'the largest exclusive motor manufacturer in the world'. It was an old-established company and its reputation would give the Ferguson tractor a great deal of

RIGHT: *A Ferguson TE-20 on test. The steel wheels were manufactured by Joseph Sankey and cost £26 a set..*

LEFT:
Harvesting with a TE-20 at Wood Farm, Ufton in Warwickshire. Wood Farm was often used by Ferguson's field-test engineers for testing tractors and implements. Note the slight kink in the exhaust pipe that was peculiar to the Continental powered tractors.

LEFT:
A Ferguson TE-20 with the Continental engine. It is seen with the manure loader and manure spreader that were added to the implement line in 1949.

RIGHT:
TE-20s on the line inside the main assembly shop at Banner Lane. The main conveyor was timed to handle 200 tractors in an eight-hour shift when on full production.

BELOW:
A diagram of the layout of the main assembly shop at Banner Lane.

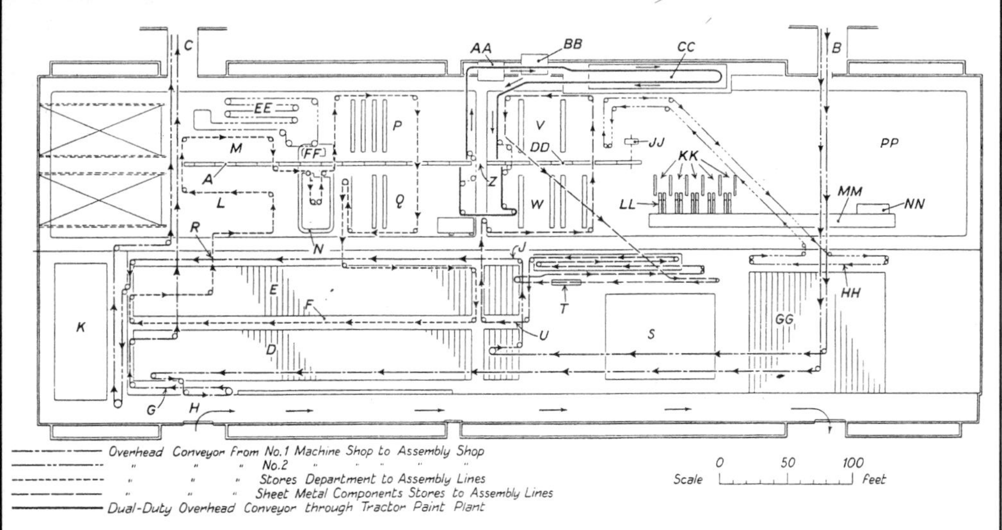

A.—Main Assembly Track. *B.*—Point at which Conveyors enter from Machine Shops. *C.*—Exit Point of Conveyors. *D* and *E.*—Roller Conveyors Serving Centre Gangway. *F.*—Centre Gangway in Stores. *G.*—Unloading Point for Conveyor from No. 1 Machine Shop. *H.*—Lorries enter Stores at this Point. *J.*—Conveyor Lowers Material to Floor Level at this Point. *K.*—Storage Racks for Bolts, Nuts, Washers, etc. *L* and *M.*—Sub Assembly Benches. *N.*—Rear Axle Assembly Track. *P* and *Q.*—Sub Assembly Sections. *R.*—Overhead Conveyor Leaves Stores at this Point. *S.*—Storage Tracks for Sheet Metal Parts. *T.*—Dip Tank for Sheet Metal Parts. *U.*—Loading Point for Electrical Equipment. *V* and *W.*—Sub Assembly Sections Serving Finishing Track. *Z.*—Gap Between Main Assembly and Finishing Tracks. *AA* and *BB.*—Paint Spray Booths. *CC.*—Drying Oven. *DD.*—Finishing Track. *EE.*—Engine Storage Conveyor. *FF.*—Engine Mounting Station. *GG.*—Gravity Roller Tracks for Wheel Storage. *HH.*—Overhead Conveyor for Wheels. *JJ.*—Wheel Fitting Station. *KK.*—Test Booths for Finished Tractors. *LL.*—Ramps for Loading Slat Conveyor. *MM.*—Slat Conveyor for Final Inspection and Adjustment. *NN.*—Ramp for Unloading Slat Conveyor. *PP.*—Parking Area for Finished Tractors

credibility on the American market – a fact that was important to Harry Ferguson. The engine chosen for the Ferguson was the four-cylinder Continental Z-120, an overhead valve petrol unit developing 23 bhp. It had an 81 mm bore, 95 mm stroke and a capacity of 1,962 cc (119.7 cu in.). The American Marvel-Schebler carburettor was retained, but its needle adjustment had to be reset to cope with the British petrol that was still affected by post-war restrictions and was of a slightly lower grade than that available in America. A British Lucas 6-volt electric system with coil ignition and electric start was also used

New features included a four-speed constant-mesh gearbox with helical-cut gears that was developed in Coventry. The Ford Ferguson had been fitted with a safety starter-button interlocked to the gear lever, but because this was protected by a Ford patent, it

LEFT:
The finishing area at the end of the main conveyor inside the Banner Lane plant. Here the tractors were furnished with petrol, oil, grease and water. Note Esso lubricants were used.

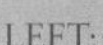

LEFT:
The Ferguson ploughs were all manufactured by Rubery Owen at a factory in Staffordshire. A conversion set was available for the two-furrow plough to turn it into a three-furrow as seen in this hand-coloured glass plate.

RIGHT: *The Ferguson steerage hoe on test in the Stoneleigh Deer Park. Colin Steventon is driving and Ted Dowdeswell is on the back.*

RIGHT: *Harry Ferguson insisted on trying every new piece of equipment for himself, and is seen at the wheel of a TE-20 using a tiller to clean the banks of the River Dikler that ran through his estate at Abbotswood. This tractor has the Lucas lighting kit that was part of the accessory range and cost £14 extra.*

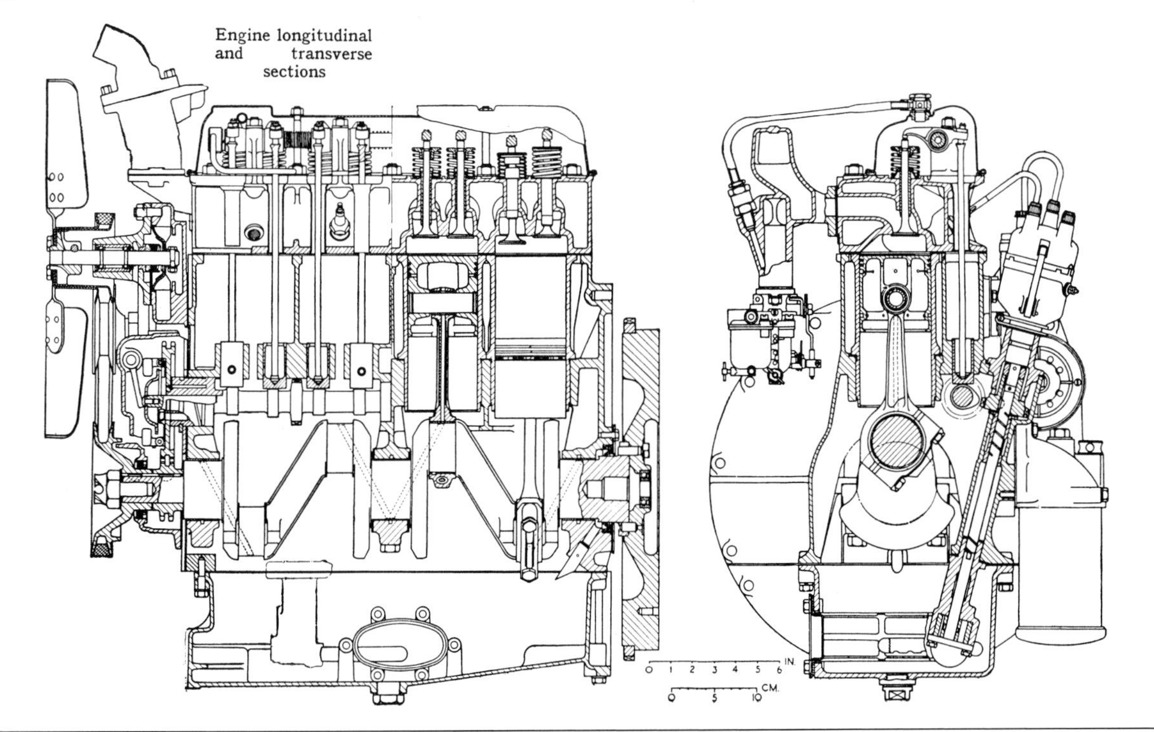

LEFT:
Sectional views of the Standard petrol engine that was introduced for the Ferguson tractor at the end of 1947. It had a three-bearing crankshaft and wet liners. Note the external oil filter.

was decided to add a fifth selector position to the gear lever to operate the starter switch.

Harry Ferguson was involved in every design change and fussed over every detail. Mindful of driver fatigue, he insisted that the clutch pedal pressure should be no more than 35 psi, and that the stalling pressure on fully locked foot brakes should not exceed 90 psi. The brakes were developed in conjunction with Girling while Borg & Beck supplied the clutch. The sheet metal styling took the longest to finalise as Harry changed his mind over it several times. The prototypes were fitted with Ford Ferguson tinwork for field trials to disguise the fact that they were new tractors on test.

In addition to the work on the tractor, there was also a range of implements to design and develop. John Chambers returned from America to assist in 1945, having appointed one of his American team, Hermann Klem, to take over his position as chief engineer of Harry Ferguson Incorporated. Harry Ferguson controlled all the

BELOW:
A Ferguson TE-A20 tractor with a two-furrow plough. Phased in during 1948, the TE-A20 was powered by the Standard petrol engine.

RIGHT:
A Ferguson TE-A20 petrol tractor working in South Lincolnshire.

designs, but often gave his engineers little guidance as to exactly what he wanted. He was known to gather Sands, Greer, Chambers and Patterson together and say, 'Now you boys know what I want; you know how I like it.'

The new tractor was designated the TE-20, the letters signifying Tractor England and the number indicating its approximate horsepower. The first was completed on 6 July 1946 and was given to Harry Ferguson. Tractor No.2 went to Sir John Black. Another twelve were then made for demonstration, sales and marketing purposes. Priced at £343, the TE-20 was launched at a special demonstration held on land adjoining the Banner Lane factory on 18 September 1946.

It was another few weeks before full production was actually underway, and only 316 tractors were built that year. Standard had completely refitted Banner Lane with machine tools obtained from the Ministry of Supply and some imported American equipment. The factory employed around 3,500 people and used the latest methods of flow production and assembly.

The main production buildings consisted of two machine shops and an assembly shop. The No.1 machine shop was used for manufacturing and machining gears, shafts and transmission components, while the No.2 shop dealt with finishing castings and producing the smaller parts. The three buildings were connected by over eleven miles of overhead roller-chain, slat and floor conveyors. The assembly shop housed the main production line as well as the paint spraying booths, and test and final inspection areas.

Production built up gradually during 1947, but was hampered by steel shortages and a fuel crisis exacerbated by one of the coldest winters for over seventy years. By May, production had risen to 100 tractors per day with one rolling off the line every five minutes. Power cuts towards the end of the year saw

BELOW:
The Standard 12cwt van was available to Ferguson dealers from January 1949 at a special price of £373. It was based on the Vanguard car that was introduced in July 1948.

the factory operating on a four-day week, but Standard's workers rose to the challenge, put in ten-hour days and built 10,000 tractors in thirteen weeks. Over 20,000 TE-20s came out of Banner Lane during 1947, and a quarter of these were exported.

Many of the tractor components were subcontracted to outside suppliers, such as castings from Frank Wade Ltd. of Stourbridge; Joseph Sankey & Sons Ltd. supplied wheel centres, sheet metal and fenders from its Hadley Castle Works in Wellington, Shropshire, while the hydraulic components came from Dowty Equipment Ltd. of Cheltenham.

The Ferguson implement range initially consisted of a two-furrow plough, a mower, earth scoop, transport box, cordwood saw, 3-ton trailer and various items of tillage equipment, including a ridger, tiller, disc harrow, spike-tooth harrow and two rowcrop cultivators. A post-hole digger, steerage hoe and spring-toothed harrow were added to the line during 1947. Although designed by Harry Ferguson Ltd., the implements were supplied by independent outside companies.

The ploughs were all manufactured by the Rubery Owen Organisation at a factory in Darlaston, Staffordshire. This plant, which employed 200 people, was set up in 1946 after visits were made to similar facilities in the USA, including Oliver in Indiana, to study manufacturing methods. All the drop-stamping,

ABOVE:
A Ferguson TE-C20 narrow petrol tractor with a 30cwt trailer. The narrow tractor was introduced in 1948 for the hop and fruit growing areas. It featured several modifications to give it a reduced width and was fitted with steel hub caps to protect the growing crop.

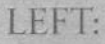

LEFT:
The Ferguson TE-D20 vaporising oil tractor that was introduced in 1949. Note the aluminium heat shield that Alex Senkowski devised to fit over the manifold and vaporise the fuel.

ABOVE:
A Ferguson TE-D20 with a 30cwt trailer. The TE-D20 model had a dual compartment fuel tank for starting on petrol and was fitted with a water temperature gauge.

RIGHT:
Ladies lend a hand to the field-test department. Nigel Liney's wife, Nan, at the wheel of a Ferguson TE-D20, while his sister-in-law is at the working end of the cordwood saw.

ABOVE:
A coloured glass plate of a Ferguson TE-D20 at work with a cordwood saw.

pressing, machining and welding processes, as well as assembly and painting, were carried out at Darlaston. The mouldboards, like all previous Ferguson ploughs, were based on an American Oliver design. A total of 17,522 ploughs were made during 1947, 20 per cent of which were exported.

Most of the tillage equipment, rigid and spring-tine cultivators, the ridger and post-hole digger were made by Steel's Engineering Products at its Crown Works on the south bank of the River Wear in Sunderland. The plant employed 1,500 people and the first implement was made in December 1946.

The trailers were all built at Sankey's Hadley Castle Works with Dowty tipping rams; John Garrington made the steerage hoe, while Thomas Blackburn & Sons Ltd. of Preston supplied the spike-tooth, spring-tooth and disc harrows. Robert Watson of Bolton in Lancashire manufactured the cordwood saw and earth scoop, and the transport box came from Gillard, Hughes & Company of Oldbury in Worcestershire. The mower was supplied from America until the Pressed Steel Company of Paisley in Scotland took over its manufacture in July 1947. A good indication of the extent to which Ferguson relied on sub-contract suppliers is illustrated by the fact that four small firms in the Midlands were noted as jointly supplying the company with a total of six million nuts and bolts a year.

There was also an extensive line of Ferguson accessories including steel wheels made by Sankey, a Lucas lighting set and a canvas cover from the Coventry Hood & Sidescreen

ABOVE: *A Ferguson TE-D20 vaporising oil tractor. The vertical exhaust pipe was another optional extra from the Ferguson accessory range.*

Company. A clever jack that relied on the tractor's hydraulics to lift all four wheels off the ground was manufactured for Ferguson by Sun Engineering of Scunthorpe.

Before any piece of machinery was put into production, it was subjected to a rigorous and intensive test programme. The foreman in charge of field test was Jack Bibby, and the field-test engineers included Dick Dowdeswell, Nigel Liney and Colin Steventon. Their brief was simple - take it out and break it. It was a test to destruction.

Any new implements were also sent to Abbotswood for Harry Ferguson's approval as he insisted on testing and trying everything. A spare seat was often fitted to the test tractors so Harry could ride with the drivers and see the equipment in action. If the machine did not live up to his exacting standards, then he refused to allow it to go into production and it was scrapped.

Ferguson's ambition was to develop a piece of machinery to suit every farming need. He wanted to offer farmers a complete system and realise his vision of taking the drudgery out of farming, to produce cheaper food and so make a better world. It was an idealistic dream, but he inspired such loyalty among his staff and engineers that they all believed in his vision and were prepared to forgive his eccentricities. They all earnestly subscribed to his motto 'Farm better, farm faster with Ferguson'.

There were several staff changes during 1947; following a disagreement over American production policies, Roger Kyes left the company and his position was taken by one of his fellow directors, Horace D'Angelo. Charles Vincent rejoined Harry Ferguson Incorporated, now based in Detroit, and Allan Botwood was appointed as the managing director of Harry Ferguson Ltd. in England. Botwood, previously a director of the Rootes Group, was a well-known figure in the Coventry motor industry; he had joined Humber Cars in 1913 and became its managing director in 1934.

Harry Ferguson preferred to surround himself with people from the motor industry rather than the agricultural machinery field as he felt they were less conservative in their approach to business, and one-third of the Ferguson dealers appointed were primarily motor traders. The distributor network, which eventually extended to well over a hundred dealers in the UK alone, was set up by Charles Turner-Hughes, an ex-British Aircraft Corporation test pilot. Sales director was another ex-Rootes man, Eric Young, while the home sales manager for the company was Trevor Knox. Ferguson's export director was Ian Wallace. One of the key personnel in the organisation was the patents manager, John Wilson, who understood the tractors and equipment inside out.

On the engineering side, Willie Sands had terminated his long-standing association with Harry Ferguson and left the company. He was not happy being away from Ulster, and it has also been suggested that he was nursing several grievances. Possibly he believed his work had been overlooked and felt that he had never been given full credit for his contribution to the development of the Ferguson System, but in truth he was 65 years old and probably felt he was ready for retirement. Archie Greer retired soon afterwards.

John Chambers was now Harry Ferguson Ltd.'s chief engineer. He was assisted by Alex Senkowski, a brilliant Polish engineer who had been a pilot in the Polish Air Force and had flown against the Russians between 1918 and

ABOVE:
The Ferguson School of Mechanised Farming at Stoneleigh Abbey. Workshop training was held in the old riding school on the right. The yard was originally a vinery. The instructor on the left in the white overalls was Ralph Herald.

LEFT:
The Ferguson lamp oil tractor, introduced in 1950, was designed to run on low or zero octane fuels for export territories. This TE-H20 model was photographed in Iran in 1951.

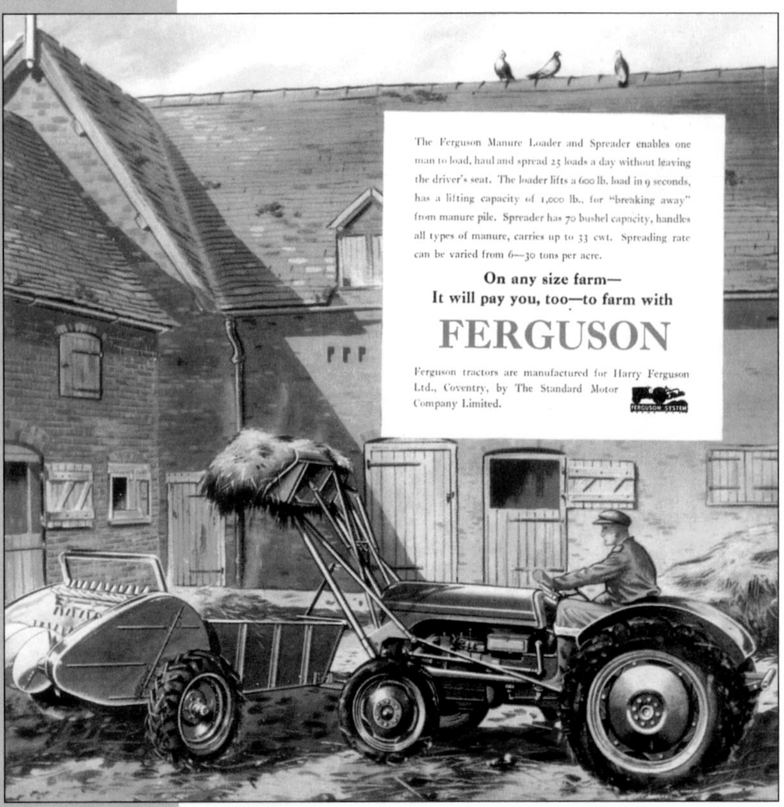

The Ferguson Manure Loader and Spreader enables one man to load, haul and spread 25 loads a day without leaving the driver's seat. The loader lifts a 600 lb. load in 9 seconds, has a lifting capacity of 1,000 lb., for "breaking away" from manure pile. Spreader has 70 bushel capacity, handles all types of manure, carries up to 33 cwt. Spreading rate can be varied from 6—30 tons per acre.
On any size farm—
It will pay you, too—to farm with
FERGUSON
Ferguson tractors are manufactured for Harry Ferguson Ltd., Coventry, by The Standard Motor Company Limited.
FERGUSON SYSTEM

ABOVE: *A 1949 Ferguson advertisement. Harry Ferguson understood the importance of sales promotion and the company had a very effective publicity department. Distributors were persuaded to plough back one per cent of their turnover into advertising.*

1920. In 1929, Senkowski became a director of the PZL aircraft factory in Warsaw. Following the German occupation in 1939, he escaped to Britain to work for the Bristol Aeroplane Company before joining Ferguson in 1946 as the tractor development engineer. He brought with him two fellow Poles: George Belkowski, a mathematician specialising in engineering stresses, and Witol Czarnocki, an engine test engineer.

By June 1947, Standard's own tractor engine was almost ready for production. It was developed entirely by Standard, and the company's technical director, Ted Grinham, was responsible for the design. At Harry Ferguson's request, it was loosely based on the Continental Z-120 with similar dimensions and layout, but various changes and improvements were made and no licensing was involved. Evidently Grinham and his team were also influenced by the design of the 1911 cc (116.6 cu in.) overhead valve engine that Citroen had introduced to power its Light Fifteen car in 1939.

RIGHT: *Test-engineer Nigel Liney discovered the limitations of this Ferguson TE-20 fitted with a Perkins P3 engine after turning it over during field trials. The three-cylinder Perkins engine was being evaluated along with Meadows and Standard diesels for possible use in Ferguson tractors.*

ABOVE:
Nigel Liney with the upturned Perkins-powered TE-20. He was lucky to escape after the tractor became unbalanced while turning at speed with a three-furrow plough and a full tank of fuel.

The Standard engine was a 1,849 cc (112.9 cu in.) unit with an 80 mm bore and a 92 mm stroke. Like the Continental unit, it had wet liners, but used a different sealing arrangement with flanged liners and 'spectacle' washers to simplify production. An external oil filter was also fitted, unlike the Continental that had a filter enclosed in the sump. It used a Zenith 24T-2 carburettor of a type originally designed for the Austin 7 car. Harry Ferguson never liked this carburettor, but it was the only option as the company had been quoted £25,000 to develop an alternative unit.

The power target for the new engine was set at 25 bhp so as not to exceed the limitations of the rear axle. To everyone's surprise, the first prototype engine developed 35 bhp. No one ever found out why, and the subsequent builds were all right on the mark at 25 bhp. The tractor engine was regulated to run at between 400 and 2,000 rpm, but Standard's engineers found that by increasing the engine speed to 4,000 rpm, they had an ideal power unit for the new Vanguard car. Whether by coincidence or design, the Standard Motor Company was able to standardise on one engine for all its tractor and car production.

Tractors fitted with the Standard petrol engine were designated TE-A20. Pre-production versions appeared from September 1947, and the first public demonstration of the TE-A20 was held on 11 December. The new Vanguard car had been announced in July, but like the TE-A20 it was waiting for Standard's new engine line at Canley to get underway. Two giant 55-ton broaching machines for the cylinder blocks were installed in the machine shops at the end of 1947, and the company announced the following January that it was nearly ready to commence production at a rate of 1,000 engines per day.

Production difficulties and material shortages saw the engine manufacturing plant get off to a slow start. The Standard-powered TE-A20 models were phased in from 26 January 1948, but tractors with Continental engines remained in production for most of the year. The price of the Ferguson tractor had risen to £372, but this was cut to £325 in February. The company claimed that this was in response to a plea from Sir Stafford Cripps for British industry to reduce

RIGHT:
A Ferguson TE-F 20 tractor with a Ferguson reversible heavy-duty disc harrow. Introduced in March 1951, the TE-F20 was fitted with the Standard 20C diesel engine designed in conjunction with Freeman Sanders.

BELOW:
Ferguson TE-F20 tractors on the production line at Banner Lane.

prices to stimulate world trade, but the more likely reason was that the use of the Standard engine reduced the manufacturing cost of the tractor.

The passing of the TE-20 with the Continental engine was not without some sadness; it had been a very dependable machine and had built up a good reputation for Ferguson tractors. There had been some minor oil consumption problems, but the engine remained a good performer, and even after the TE-A20 was launched, it was the Continental tractor that was always chosen for demonstrations against the Fordson Major.

Several modifications were made to the Standard engine during 1948, including different steel for the exhaust valves and an inclined Tecalemit oil filter to replace the earlier vertical type. An enlarged 2,088 cc (127.4 cu in.) version of the engine with an 85 mm bore was also introduced for the Vanguard.

Vanguard production began in July 1948, but only just over 2,000 had been made by September and most of these were sold for export. The same month, Standard announced estate, van and pick-up versions. The 12 cwt delivery van was offered to Ferguson distributors and dealers at a special concessionary price of £373 from January 1949, and it became the usual sales and service vehicle, replacing the Standard Fourteen vans that had been used previously.

LEFT:
A 1953 Ferguson TE-F20 diesel tractor that was supplied by the South Lincolnshire dealers, Boston Tractors.

The Ferguson TE-20 tractor passed the production milestone of 50,000 in August 1948, but demand, particularly for export, was still exceeding supply. In an attempt to ease the steel shortage, Harry Ferguson initiated a nationwide scrap metal drive to persuade farmers to part with their obsolete equipment.

In September 1948, the company announced a narrow version of the Ferguson tractor with the wheel-track reduced from 52 to 42 in. Designed for the hop and fruit growing areas of the country, it had shortened rear-axle housings and shafts as well as modifications to the front axle and radius arms. Alterations also had to be made to the linkage, and included cranked lift arms, shortened check-chains and a sliding bar in place of the winder to operate the levelling gear. The brake pedals also had to be redesigned, and steel hub caps were fitted to protect the growing crops from damage from the front-wheel hubs and nuts.

A few components for narrow tractors for the French market had been manufactured 'in house' at Banner Lane, but the modifications for the narrow tractors to be sold in Britain were carried out on behalf of Harry Ferguson Ltd. by the Lenfield Engineering Company. Lenfield, who were Ferguson dealers with three depots in Kent, modified the tractors in Sittingbourne using castings and forgings supplied by a subsidiary firm, Wellwinch Engineering. The very few narrow tractors that had the Continental engine were designated TE-B20, while the Standard-powered machines were known as the TE-C20 and were priced at £360.

In 1949, the government finished the so-called 'red petrol' scheme that allowed the fuel to be rebated for agricultural use. Faced with 9d a gallon tax on petrol, the Ferguson company rushed to introduce a tractor to run on a cheaper distillate fuel. The new vaporising oil model was designed using Standard's larger 85 mm bore engine to compensate for the loss of efficiency resulting from the lower-grade fuel. The cylinder head was modified and the compression ratio was lowered to 4.8:1. A different type of five-ring piston was used to minimise wear and oil dilution.

Alex Senkowski devised an aluminium heat shield to fit over the manifold to vaporise the fuel. It was a simple arrangement, but Alex Patterson still claims that it was one of the best vaporisers in the world. The thermostat settings were changed and a water-temperature gauge

RIGHT:
The Ferguson TE-F20 tractor. Note how the batteries have been relocated to the side of the seat, as there was no room under the hood because of the increased dimensions of the diesel engine and its extra ancillaries.

BELOW:
A version of the 20C diesel engine as used in the Ferguson tractor was fitted to the Standard Vanguard car from 1954. This 1955 example is the Vanguard Phase II model.

was fitted to the dashboard. To enable the tractor to be started on petrol and then changed over to vaporising oil once warm, it had a dual compartment fuel tank with a three-way fuel tap.

In the panic to get the vaporising oil model on the market as soon as possible, an intensive test programme was implemented with the tractors working twenty-four hours a day through the early months of 1949. The field-test engineers worked twelve hour shifts, ploughing during the day and running on the road with a fully-laden 3-ton trailer at night.

The TE-D20 vaporising oil tractor was unveiled at the Royal Dublin Show in May and was put into production in July. It was rated at 26 bhp and cost £335. The narrow version was known as the TE-E20. The Ferguson tractors also benefited from lift improvements introduced in January 1949, with increased hydraulic pressure and strengthened components.

The 100,000th TE-20 tractor was completed on 13 June 1949. Steel shortages had eased and production was approaching the level of 300 per day. The 'little grey Fergie', as it was affectionately dubbed, had built itself an enviable reputation in a very short time and had made an impressive impact on the market. The success of the tractor was due not only

to the qualities of the machine itself, but also to the well-organised dealer network with considerable back-up from the company and well-prepared demonstrations.

Harry Ferguson Ltd. ran its own sales and service training schools to train dealer personnel. The first of these was set up at Fletchamstead in early 1946, later moving to Packington Hall near Stonebridge on the outskirts of Birmingham. In 1949, the Ferguson School of Mechanised Farming was established at Stoneleigh Abbey after the company negotiated an agreement to rent some of the buildings from Lord Leigh. The old riding school and stables were set up as workshops and part of the Abbey building was used as a residential hall for the students. Around 150 acres in an area known as the Deer Park were taken over for field operations. The school was run by John Chambers' brother, Dick, and as the demand for courses increased, the dormitory, lecture rooms and workshops were moved into an old American camp on the estate. A similar establishment was also opened in the Republic of Ireland at Powerscourt in County Wicklow.

Overseas orders for the Ferguson tractor were also flowing in, and around 12,000 were exported in the first six months of 1949. The vaporising oil tractors were designed to run on 50 octane fuel, but in some countries this was not available and farmers had to rely on fuels

ABOVE: *The vineyard model Ferguson that was launched in 1952 for wine and sugar cane growing areas requiring an extra-narrow tractor. Smaller wheels gave it a lower profile and reduced the centre of gravity.*

LEFT: *Ferguson industrial tractors, designed for municipal use, general haulage and building and construction work, were introduced in 1951. This TE-P20 petrol model has the full-width fenders and the dual braking system, and is seen fitted with a Hydrovane 60 compressor.*

RIGHT: *A Ferguson TE-T20 diesel industrial tractor. Seen with a hedge-trimmer attachment, it has the spring-loaded bumper for protection or shunting, and heavy-duty industrial tyres.*

with an even lower octane rating or often unknown specifications. To meet those requirements, Ferguson introduced the zero octane or lamp oil tractor for export only from April 1950.

The TE-H20 normal-width and TE-J20 narrow lamp oil tractors were based on the vaporising oil models with a lower 4.5:1 compression ratio and retarded ignition to prevent pinking. They were under-powered and only developed 22 bhp, and there were some problems with carbon deposits forming on the pistons, which were not helped by the lack of detergent oils. However, the tractors would run happily on Esso Blue paraffin, or similar. Ferguson's field-test engineers found that if you got the lamp oil engine hot enough it would even run on diesel, although you could not see for smoke!

The Ferguson implement range was continually updated and extended. Seven new machines were announced on 18 January 1949, including a manure loader and a potato spinner from Steel's Engineering Products; a new

RIGHT: *A TE-PT20 semi-industrial petrol tractor with a high-lift loader in the engineering works at Fletchamstead. It has full rear fenders, bumper and the road lighting kit.*

potato planter was manufactured by Midland Industries Ltd. of Wolverhampton, and William Donaldson of Paisley supplied the 'Skidmaster' wheel girdles. The other additions were a subsoiler, weeder and manure spreader.

An earth-leveller or blade-terracer and a disc plough appeared in February 1950, with a linkage winch, manufactured by Hesford of Ormskirk in Lancashire, following five months later. The same year, a mounted hammer-mill made by Scottish Mechanical Light Industries of Ayr was incorporated into the range. An impressive number of around 350,000 Ferguson implements had been sold worldwide by the end of 1950.

Several changes were also made to the tractor range. For a short time, from March 1949, a few petrol engines were fitted into Holley carburettors. An oscillating hydraulic control valve was introduced in June 1950, and in February 1951 Standard dropped the old 80 mm bore petrol engine to make room at Canley for a new diesel. This meant that the TE-A20 had to be fitted with the same 85 mm bore unit as the vaporising and lamp oil tractors and now developed 28.4 bhp. The following month, after Lucas announced that it was discontinuing its 6-volt electrical systems, the TE-20 bell-housing was altered to accommodate a different starter motor as the

ABOVE:
Many Ferguson tractors were used by local authorities for light haulage and road or playing field maintenance. This TE-T20 diesel model was used for municipal work in Surrey and is seen with an industrial version of the 3-ton trailer with a sprung axle and over-run brakes.

LEFT:
A 1954 Ferguson TE-PT20 semi-industrial petrol tractor with a front-mounted Sturdiluxe brush attachment. It spent its working life sweeping roads at a quarry in Leicestershire.

ABOVE: *The high-lift loader was one of seven new items of Ferguson equipment launched during 1951. Ferguson's engineers referred to it as the cantilever loader and it is seen fitted to a TE-D20 outside the engineering works at Fletchampstead.*

changeover was made to 12-volt electrics.

A diesel-powered Ferguson tractor was a new departure for the company, but one that had been in the planning stage for some time. The Standard Motor Company had developed its own compression-ignition engine in conjunction with the Freeman Sanders Engineering Company based in Penzance, Cornwall. Arthur Freeman Sanders was a leading combustion engineer who had been employed by John Fowler of Leeds where he had designed both agricultural and industrial diesel engines.

Standard's diesel engine was a 2,092 cc (127.7 cu in.) unit with an 80.96 mm bore and a 101.6 mm stroke. Although its dimensions were only slightly greater than those of its spark-ignition counterpart, it was a much heavier power unit. The inherent design and operating characteristics of the diesel engine with its higher 17:1 compression ratio meant that it had to be manufactured from more robust components. The extra diesel injection ancillaries and heavy-duty electrical equipment all added to the weight, which was around 2 cwt more than the petrol engine.

Freeman Sanders used an indirect-injection design with a cylinder head incorporating his patented spherical combustion chambers. CAV Ltd., part of the Lucas Organisation, supplied the BPE-type in-line fuel injection pump, which was pneumatically governed. Unlike the spark-ignition engines, the diesel had dry liners with special cuff-rings at the top of the bores.

The decision to begin developing a diesel engine was made by Ted Grinham in the hope that it would be adopted for use in the TE-20 tractor. Harry Ferguson was not consulted and his engineers had had no input on the original design, so there was no guarantee that this would happen. Ferguson was certainly no advocate of diesel engines. Apart from the fact that they were heavy, complicated and more costly to manufacture, he abhorred them as dirty and noisy.

One of Harry Ferguson's favourite challenges to his engineers was 'Why can't my tractor be as quiet as my Rolls-Royce?' He was always complaining about the efficiency of the silencers, and both Burgess and Tyzack mufflers were tried. He disliked the upright exhaust, and the position of the downswept pipe left no room to make the silencer any larger, so the engineers' only option was to make it longer.

The Ferguson company, and in particular the sales department, was in little doubt that it had to add a diesel model to its range; other manufacturers had diesel tractors in the pipeline, and farmers were attracted by their reliability, economy and the fact that they ran on cheaper fuel. In some ways, Standard had pre-empted Ferguson's intentions, but the latter

ABOVE:
A TE-D20 tractor at work with a high-lift loader, often dubbed the 'banana' loader because of its distinctive shape.

company resolved first to evaluate other suitable diesel engines before making a decision.

Two other diesel power units were fitted to TE-20 tractors to be pitted against the Standard engine, the three-cylinder Perkins P3 and a four-cylinder Meadows. The Meadows performed best and had excellent torque, but was a poor starter and was expensive. The Perkins P3 was a good engine, but to accommodate it the TE-20's fuel tank and hood line was raised, which Harry Ferguson felt was not aesthetically pleasing to the eye. Furthermore, this gave the tractor a high centre of gravity, as confirmed by one of the field-test engineers, Nigel Liney, who managed to tip a Perkins-engined TE-20 completely upside down when executing a quick turn on independent brakes with a full tank of fuel. Miraculously, he was uninjured.

In the end, the Standard diesel, known as the 20C, was the only economical choice. It had its drawbacks and could be slow to start when cold, but to overcome this it was fitted with a Ki-gass pump to spray atomised fuel on to the glowing filament of an induction heater in the inlet manifold. The intensive test programme also uncovered a worrying tendency for the engine to run backwards after one of the tractors was nearly stalled on a hillside. A shocked Dick Dowdeswell found himself hurtling back down the slope in reverse with flames coming out of the air breather as the engine exhausted through the inlet manifold. Small modifications were made to the injector pump to ensure that this was unlikely to happen again.

The Ferguson TE-F20 diesel model was introduced in March 1951 and cost £490. Outwardly, it was almost identical to the other tractors in the range except that the two 6-volt

RIGHT: *The Ferguson buckrake behind a TE-D20 tractor. Designed for handling hay and green crops, the buckrake was one of about six new Ferguson implements brought out in 1952.*

batteries had to be relocated either side of the seat as there was no room under the hood. Slight alterations also had to be made to the transmission to accept the new engine.

The TE-F20 was one of the first British diesel tractors to go into volume production, and the first example to be sold, tractor No. 200033, went to a local farmer, Jim Russell of Sandpit Farm near Rugby. Flushed with the success of the 20C, Standard were to offer a version of the engine for the Standard Vanguard car and commercial vehicles three years later. The diesel Vanguard, launched in February 1954, was the first British production car to be fitted with a compression-ignition engine and was capable of returning over fifty miles to the gallon.

The development programme for the TE-F20 alone saw the test tractors clock up several thousand hours either in the field or on the dynamometer in the workshop. The agenda for testing Ferguson equipment was ongoing and intensive with both tractors and implements to be evaluated. Ferguson's engineers also explored the possibilities of using a unique rotary combustion cell in the spark-ignition engines.

This rotary-valve cylinder head was developed by F. M. Aspin from Bury in Lancashire. Harry Ferguson had shown interest in the design and had added the Aspin company to the portfolio of Ferguson Holdings. Instead of the usual cylinder head with poppet valves and tappets, Aspin substituted a cone-shaped rotor that was inserted over each cylinder and driven off the crankshaft. This was claimed to make the engine more efficient and allow a higher compression ratio to be used with low octane fuels. It would have been a costly design to manufacture, and as Ferguson's field-test engineers reported no marked improvement in performance, no attempt was made to put it into production.

As the narrow-width tractor was still too wide for certain applications, Ferguson introduced a vineyard model in May 1952. Designed for wine and sugar cane growing areas, it was more extensively modified than the narrow model and was only 46 in. wide with a wheel-track of 37 in. It had similar shortened half-shafts and revised lift linkage, but to accommodate the alterations to the front axle and radius arms, an extra support casting was added to move the radiator forward which meant that the hood had to be lengthened by 4 in.

The vineyard tractor also had smaller 24 in. rear and 15 in. front tyres that lowered its overall height by some 2 in. This was both to give it a reduced profile for orchard work and to keep its centre of gravity low. Three engine options were available: the TE-K20 was the petrol model, while vaporising oil and lamp oil vineyards were designated TE-L20 and TE-M20 respectively. Narrow and vineyard versions of the diesel tractor were only produced in France.

Not content with its success on the agricultural market, Ferguson widened the scope of the TE-20 to attract sales to the building and construction trades and municipal corporations by introducing the TE-P20 petrol industrial model in April 1951. The tractor conformed to the legal requirements for public road work and had two independent braking systems: hydraulically operated foot brakes and a mechanical handbrake. It was equipped with detachable full-width fenders, industrial tyres, a horn, rear-view mirror, and a spring-loaded bumper for protection or shunting.

The Ferguson industrial tractor proved popular with councils and local authorities for general transport, refuse collection and road, park and playing field maintenance. They were ideal for light haulage, and were extensively used in factories, docks and timber yards. The TE-R20 vaporising oil and TE-S20 lamp oil industrial tractors were added to the range in 1952.

After the TE-T20 diesel industrial tractor was launched in February 1953, the vaporising oil and lamp oil industrials were slowly phased out. Other changes saw the industrial models referred to as Ferguson mobile power units to emphasise their versatility. There was a greater choice of specification and the customer could choose from a range of accessories to tailor the tractor to his needs, from a basic model with just dual braking, industrial tyres, a horn and mirror, to a full-industrial with all the equipment.

BELOW:
The Ferguson low-volume sprayer at Stoneleigh. Launched at the 1952 Smithfield Show, the sprayer was manufactured for Ferguson by Fisons Pest Control of Cambridge.

ABOVE:
A 1953 Ferguson TE-D20 on test with a two-furrow plough near Coventry. The driver is seen operating the furrow-width adjuster that became part of the accessory range.

To confuse matters, as further options and the choice of industrial models grew, an extra 'T' suffix was introduced for the semi-industrial version. This was followed in 1955 with a 'ZE' suffix designating agricultural fenders and a 'ZD' suffix for industrial tractors supplied without fenders.

Ferguson did not really supply any industrial equipment as part of its range, although several implements, such as trailers, loaders, the subsoiler, post-hole digger and earth-leveller, were adapted for industrial use. The exceptions were probably the Hydrovane compressor and the earthmover that were added to the implement line in 1954. Several industrial machines to suit the TE-20 were marketed by other manufacturers, including the Sturdiluxe brush attachment, Tayman pole-hole borer, Allen Atos trench-digger and the Modern Plant Autoloader for moving ballast. Twose of Tiverton offered hedge and verge mowers and the Tractamount roller attachment.

The range of 'badged' Ferguson equipment continued to grow. The first half of 1951 saw seven new introductions: three-furrow conventional and single-furrow reversible ploughs, a mounted offset disc harrow, high-lift loader, hay sweep, precision seeder and a combined seed and fertiliser drill. The precision seeder was designed for the export market and was capable of sowing maize, cotton, peas or beans. The seed and fertiliser drill was a development of the universal seed drill that had been put into production the previous October.

The Ferguson seed drill was designed 'in-house' by Ron Sargeant using a fluted-roller type seeding mechanism developed in America by Dempster Industries of Nebraska. He was assisted by Stan Hockey who joined Ferguson from the British seed drill manufacturers, Dennings of Chard. Sargeant was also responsible for the high-lift loader. Ferguson's engineers referred to this as the cantilever loader, but it became better known as the 'banana' loader because of its distinctive shape. The product engineer for the loader was Theo Sherwen who had designed the 3-ton trailer and an automatic hitch attachment.

ABOVE: *The Standard 6 cwt van, announced in 1954, was an economical alternative to the Vanguard for the smaller Ferguson dealers.*

More machines followed in 1952, including the independent-gang steerage hoe, mounted tandem disc harrow, 30-cwt trailer and the buckrake. The same year saw the Ferguson game flusher exhibited at the Royal and Royal Highland shows. This consisted of a tractor-mounted boom from which weighted chains were suspended. It passed through the crop in front of the mower cutterbar to flush out pheasants and partridges to protect the young game birds. The Ferguson low-volume sprayer, manufactured for the company by Fisons Pest Control of Cambridge, was launched at the Smithfield Show in December, and the single-furrow plough came out the following year.

By the end of 1952, sales of Ferguson equipment were booming. Over 302,000 TE-20 tractors had left Banner Lane, and the number of implements produced was fast approaching the half-million mark. This kept the subcontract firms busy, and Steel's Engineering Products at Sunderland were building over 50,000 Ferguson machines a year. Harry Ferguson, now aged sixty-eight, should have been a satisfied man, particularly as his court case with Ford was finally settled and he was over £3.3 million richer. But he remained characteristically restless even though he had achieved many of his lifetime's ambitions and the Ferguson System had become accepted almost worldwide.

In fact, Ferguson was becoming slightly disillusioned with the farm machinery business, and was disappointed that he had been unable to reduce the prices of his tractors and equipment low enough to mechanise many of the developing countries. Rising production costs meant that Ferguson tractors were now even dearer than ever. By February 1953, the petrol TE-A20 cost £395 while £525 was needed to buy the TE-F20 diesel model. The American concern, Harry Ferguson Incorporated, had been struggling for several years and was becoming a drain on his resources. With no son to take over from him, and other interests in the motor industry to consider, Harry Ferguson felt it was time to contemplate a merger and opened talks with the Canadian Massey-Harris Company at the end of 1952. By the following summer, the negotiations were complete and Massey-Harris-Ferguson was formed from an amalgamation of the two firms. Ferguson tractors were now under new ownership and changes were on the horizon.

CHAPTER 5

Ferguson Overseas

A Ferguson TE-F20 diesel tractor working with a reversible heavy-duty disc harrow on a banana plantation in the antipodes. The tractor was probably supplied through New Zealand as TE-20s sold in that country had a blue stripe around the hood.

Ferguson

RIGHT: *This TE-F20 diesel tractor being used for road construction in Malaysia was one of over 300,000 Fergusons that were exported from Banner Lane to all parts of the world.*

In addition to its success on the home market, the Ferguson TE-20 was also exported in great numbers to many parts of the world and satellite production was established in several countries. By March 1956, the 'little grey Fergie' had been on the market for over nine years and was nearing the end of its production run. However, sales remained buoyant and that month saw the 500,000th TE-20 leave Banner Lane. Out of this total of half a million tractors, no less than 318,000 had been exported to 117 different countries.

Scandinavia had become the first and remained the largest export market for the TE-20. Twenty Fergusons were exported to Norway in 1947, followed by ten each to Sweden and Denmark. By the time TE-20 production came to an end in 1956, Scandinavia accounted for 100,000 tractors with over 40,000 in Denmark, 16,000 in Norway, 10,000 in Finland and 1,000 in Iceland.

Australia was the second largest market for Ferguson tractors and 51,075 were sold there between 1949 and 1956. The main distributor was British Farm Equipment, a Standard Motor Company subsidiary with bases at Sydney in New South Wales and Melbourne in Victoria. The other states were covered by sub-distributors appointed by British Farm Equipment, such as the Tractor and Implements Company in Queensland.

The tractors were shipped from Banner Lane to Australia in part-knocked-down (PKD) form and then reassembled in Sydney or Melbourne. The only real modification made for the Australian market was the replacement of the downswept exhaust systems with vertical silencers to eliminate the risk of bush fires. Wider rims were usually fitted in place of the standard 19 in. front wheels, and the front tyres were often water-ballasted to add much-needed front-end weight to enable the tractors to cope with heavier implements. The petrol model was the most popular version of the Ferguson in Australia, followed by the TE-F20 diesel tractor. Most of the vaporising oil TE-D20 models sold to the continent were converted back to straight petrol.

The majority of Ferguson sales in Australia were made to the fertile river lands and coastal fringes of Queensland, New South Wales, Victoria and South Australia as the tractors were used predominantly on arable farms and tropical plantations. The TE-20 was also popular on the large sheep and cattle stations in the inland for fencing and other jobs, where they were used as 'packhorses', travelling around the property with trailers or transport boxes.

Outside of Europe, the other main markets

ABOVE:
Ferguson tractors and equipment on demonstration in Australia at a field day held near Camden in New South Wales during the 1950s. The TE-20 nearest the camera is working with the subsoiler that was introduced in 1949.

LEFT:
A Ferguson TE-20 receives an on-farm service visit in Australia. The Standard Vanguard service van belonged to the main distributors, British Farm Equipment.

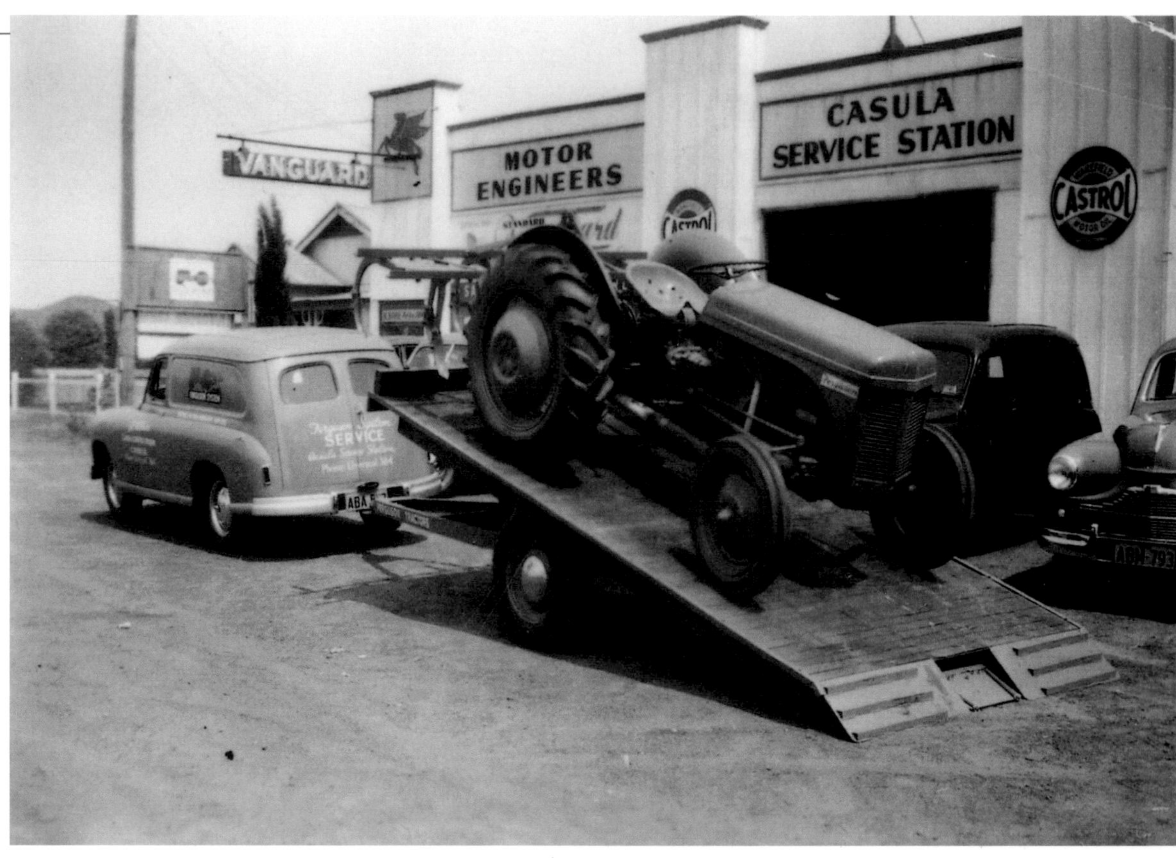

RIGHT: *A TE-D20 tractor and tiller loaded on a transporter trailer at a dealership in Australia. The dealer, R.H. Bare, was based in New South Wales and also held the franchise for Standard-Triumph cars.*

were South Africa with 21,040 tractors and New Zealand with 17,190. At the other end of the scale, 618 Fergusons were sold to Fiji, while 54 machines went to the Leeward Islands. South Africa and the Antipodean markets of Australia and New Zealand were actively developed to stimulate a demand for tractors during the slack winter months at Banner Lane.

Turkish farmers were also keen customers for Ferguson equipment, encouraged by long-term low-interest loans from the Turkish Agricultural Bank. The bank would advance 75 per cent of the value of the tractor on credit provided that at least three implements were bought at the same time. It was a situation that the Ferguson organisation was able

RIGHT: *Ferguson tractors were very popular in Turkey, and over 8,000 had been imported into the country by 1956. Ferguson's Middle-East representative, Roy Harriman, puts a TE-H20 lamp oil model through its paces at Anatolia in 1952.*

LEFT:
Ferguson tractors and Standard vehicles in the workshop of the main dealer, Rasit Ener, at Adana in Turkey in 1952.

to exploit, offering a complete tractor and implement package through its Turkish distributors, the British Overseas Engineering & Construction Corporation. In 1950, only 916 Fergusons were sold into Turkey, but the company quickly dominated the market and was supplying over 60 per cent of the tractors brought into the country from the UK. The 5,000th TE-20 was delivered to Turkey in February 1952, and by 1956 over 8,000 Fergusons had been imported.

Ferguson tractors were also exported as far afield as South America, Iran, Aden, Syria, Somalia, Egypt and the Lebanon. Six were sold to Afghanistan in 1950, and 1,000 were exported to Israel, mainly as gifts donated by Jewish benefactors in America. March 1954 saw the first full-scale demonstration of TE-20s in Ethiopia. The demonstration was held on the Imperial Farm at Addis Ababa where the Emperor Haile Selassie insisted on inspecting the tractors and implements personally before declaring them to be 'easier to handle than a team of oxen'.

Twenty Ferguson tractors were sent for evaluation in China, but the Chinese literally tested them to destruction and succeeded in breaking twenty crankshafts. Even so, the company was eventually awarded a contract in 1953 to supply a considerable order to the

BELOW:
A Ferguson field course held in Anatolia, Turkey in 1953. Roy Harriman instructs local farmers on the basic maintenance of a TE-H20 tractor.

RIGHT:
A Ferguson tractor on demonstration in Ethiopia in 1954.

BELOW:
A Ferguson tractor and disc plough in Iran in 1951. The wheel girdles were fitted to increase grip to cope with the hard and dry conditions.

republic. The TE-20 got a better reception in Japan following a successful demonstration staged at the Japanese Ministry of Agriculture and Forestry experimental station in April 1954. Up until then, Japan had tended to shun farm mechanisation, believing it to be unsuitable for its labour-intensive type of agriculture. Ferguson claimed that the TE-20 was the first tractor to really impress the Japanese. Interest was shown from the large-scale farms in Hokkaido and Honshu, and the grey Ferguson became one of the first western tractors to be imported into the territory, albeit in very small numbers.

Most of the TE-20s were exported in PKD form with two or three tractors crated in each packing case. The design of the packing case was based on the crates used to ship Ford Ferguson tractors into the UK during the war. The overseas consignments were transported by road to London and shipped from the Royal Albert or

LEFT:
Part-knocked-down TE-A20 tractors being crated for export at Banner Lane in 1948. Two tractors were packed into each crate ready for shipment from London Docks.

King George V docks. Ferguson's export successes were matched by those of the Standard Motor Company which sent 14,500 motor vehicles abroad during the first six months of 1949 to join the 12,000 TE-20 tractors exported during the same period.

To maintain a steady supply of allied equipment for the worldwide markets and overcome the steel shortages, Ferguson decentralised the production of certain implements, awarding contracts to several overseas manufacturers. D.J.J.Bekker of Johannesburg produced Ferguson disc ploughs and disc harrows for South Africa, while Aktiebolaget Overums Bruk of Sweden manufactured Ferguson ploughs for Scandinavia. Ferguson tillers were also made by S.A.C.A. of Seville in Spain and a rice transplanter was built in Italy.

The European market for Ferguson tractors was well established. Figures show that by 1956, 5,200 TE-20s had been supplied to Belgium, with 5,800 going to Holland and 8,000 to Italy. Although no numbers were available for Switzerland and Germany, 713 were shown as being sold in the Channel Islands. France was the main European and third most important overseas market for the TE-20 with satellite production facilities established there in 1953.

Demand for Ferguson tractors in France was almost immediate, particularly for vineyard models, and an order of 5,000 TE-20s was received from the country as early as February

BELOW:
Ferguson tractors on the final assembly line inside the St. Denis plant.

ABOVE: *French Ferguson tractors lined up in the dispatch yard outside the St. Denis plant.*

1948. Like Britain, France faced a postwar shortage of farm machinery. However, it had an arable area three times that of the UK's to till and needed an estimated one million tractors to mobilise its farms. Ford Ferguson tractors were imported immediately after the war and proved popular. British-built Ferguson TE-20s were an obvious choice for the country and the Compagnie Générale du Machinisme Agricole (COGEMA) was appointed as the French distributor in 1947. However, France had devalued the franc and currency problems saw heavy restrictions being placed on the import of manufactured goods making trade with Britain difficult, and COGEMA was allocated import licences for only 1,000 TE-A20 models during 1948.

The solution was for Ferguson to open a tractor assembly plant in France. Negotiations during 1952 between the French government, Harry Ferguson Ltd. and the Standard Motor Company led to a factory being established at St. Denis on the outskirts of Paris. Standard set up a subsidiary company, Société Standard-Hotchkiss in affiliation with an established French car maker to handle the manufacturing of the tractors. The COGEMA concern was also taken over and reorganised as a new sales company known as Harry Ferguson de France, run by Jacques Boilliant-Linet, becoming the

RIGHT: *A French-built Ferguson TE-K20 vigneron tractor that was assembled at St. Denis near Paris by Standard-Hotchkiss from 1953.*

Compagnie Massey-Harris-Ferguson after the 1953 merger.

Tractor assembly at the St. Denis plant began in 1953 using components exported from Coventry. The French government agreed to a three-year importation plan during which time the British parts would be slowly phased out and replaced by locally-sourced French components. Production began slowly with 2,500 tractors made the first year, doubling to 5,054 for 1954. By 1955, the annual production figure had risen to 13,794 with the engines now manufactured by an associate company, Société Hotchkiss-Delahaye.

The French Ferguson tractor range consisted of the TE-A20 petrol model, with its TE-C20 étroit (narrow) and TE-K20 vigneron (vineyard) versions, and the TE-F20 diesel tractor. Two further models that were unique to the French market, the diesel-powered TE-G20 narrow and TE-N20 vineyard tractors, were later additions to the line-up.

There were 15,922 Ferguson tractors made in France during 1956 as they rolled out of St. Denis at a rate of more than 100 a day. The company was now the most prolific tractor manufacturer in France, even eclipsing the output of the country's national marque, Renault. In September 1956, the French TE-20 was revamped and re-launched as the Ferguson FF-30. It was

ABOVE:
A 1955 sales brochure for the French Ferguson tractors from St. Denis.

LEFT:
A Ferguson FF-30DS, a normal-width diesel tractor. Introduced in September 1956, the French FF-30 range was painted red and grey to reflect the impending amalgamation of the Massey-Harris and Ferguson lines. Note the extra passenger seat on the left-hand fender

RIGHT: *India was an important market for Ferguson tractors, and sales offices, an assembly plant and training schools were established in the country. A group of trainees on a Ferguson field course is seen crossing a river in southern India in 1951.*

ABOVE: *A Ferguson tractor and plough working in a tea garden in Assam. The tractor is preparing land for the planting of legumes to provide nitrogen for a following crop of tea bushes.*

basically the same machine except that slight tuning modifications coaxed 30 hp out of both the 85 mm petrol and 20C diesel engines.

There were six variations of the FF-30 with standard, narrow and vineyard versions of both the petrol and diesel models. The FF-30 designations were suffixed D for diesel or G for petrol, followed by S for standard, E for étroit or V for vigneron. For example, the narrow petrol tractor was known as the FF-30GE. A colour change to red chassis and grey sheet metal and wheels was also introduced with the launch of the new models. It was the first attempt at amalgamating Massey-Harris red with Ferguson grey.

In April 1957, the Société Standard-Hotchkiss announced that the FF-30 was now 100 per cent of French manufacture. That year, the company produced nearly 20,000 tractors. The St. Denis plant had been extended, but further expansion was restricted as it was French government policy to limit industry in the Paris environs. It was decided to establish a new factory on a 45-acre site at Beauvais, north of Paris on the road to Calais. Beauvais was only forty miles from St. Denis and had both rail and air links.

Initially the new plant was only used for work previously carried out by subcontractors, such as machining, but in 1958 St. Denis was closed and Beauvais was expanded to take over tractor production. The FF-30 line was also dropped in 1958 and was replaced by the Massey Ferguson 835 that was launched from Beauvais in June. The 835 was the French variation of the Massey Ferguson 35 and was fitted with a Hotchkiss-Delahaye version of the Standard 23C diesel engine.

Another satellite plant for Ferguson tractors was established in India. Again, the manufacturing side was set up by the Standard Motor Company through one of its subsidiaries, Standard Motor Products of India Ltd. The factory in Madras opened in 1950 and assembled Ferguson tractors shipped from Banner Lane in completely-knocked-down (CKD) form.

The first Ferguson tractors had been sent to India in 1948, and were imported at the request of the country's prime minister, Pandit Nehru, following the advice of his sister, the Indian High Commissioner in London, who had attended a demonstration of the TE-20. The Indian government was keen to mechanise its country's farming and saw the light and uncomplicated Ferguson tractor as the ideal tool for the job.

Ferguson established its first base in the country in 1949 when it set up a sales company, Harry Ferguson (India) Ltd., headed by Roland Heath with offices in Bangalore in the province of Mysore. Two distributorships were appointed: Komani, operating out of Bombay, handled sales for the south of the country, while Escorts, under Harry Nanda and based in Delhi, were distributors for northern India.

As sales increased, training schools were established in Bangalore, Delhi and Bombay, and an assembly plant was set up in Madras. Initially, the only local content that the tractors had were the tyres that were supplied by Goodyear of India. However, following government pressure, some other components, including batteries and electrical equipment, were later sourced locally.

Some Ferguson equipment, including 3-ton

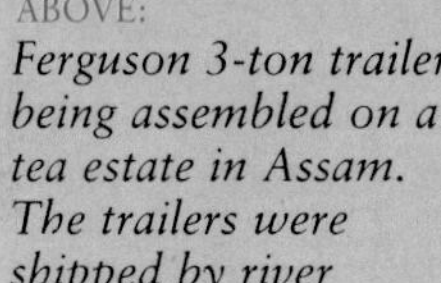

ABOVE:
Ferguson 3-ton trailers being assembled on a tea estate in Assam. The trailers were shipped by river steamer from Calcutta.

LEFT:
A demonstration of Ferguson FE-35 tractor and FE-93 ploughs in India, organised by the Delhi distributors, Escorts. The tractor in the background is an older TE-20 model with its air-cleaner intake modified to cope with dusty conditions.

V 114
FILE COPY

March 24, 1953 E. V. BUNTING 2,632,628
POWER LIFT FOR TRACTORS
Filed Nov. 25, 1946 4 Sheets-Sheet 1

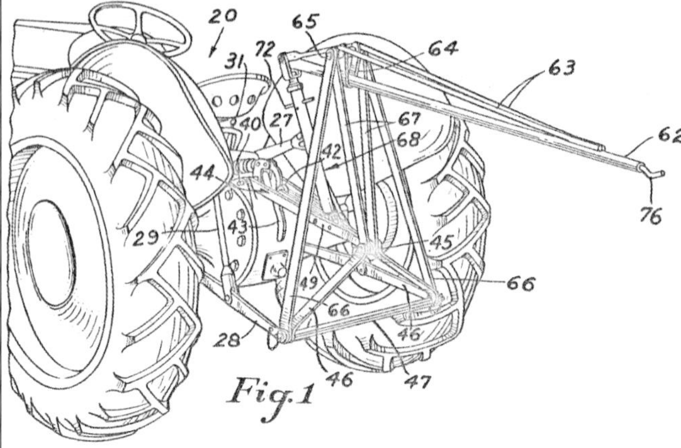

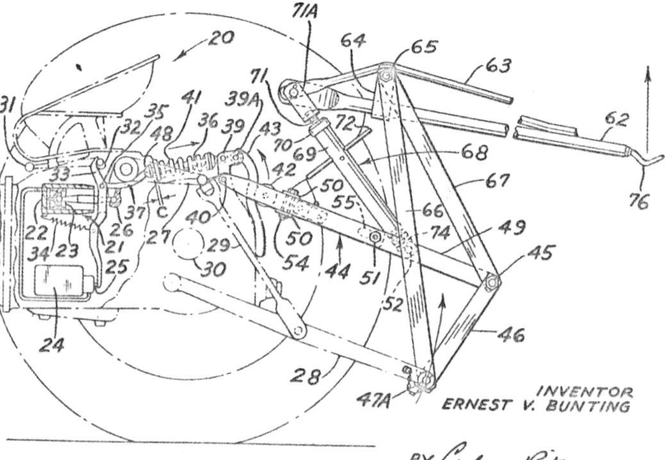

RIGHT: *Detail of Ernest Bunting's patent for a position control system and crane attachment that was filed in the USA by Harry Ferguson Incorporated in November 1946. The roller and cam lever arrangement for the position control is indicated by the number 42 on the drawings.*

trailers, transport boxes and irrigation pumps, were manufactured in India by Balmer Lawrie Ltd. of Calcutta. This manufacturing agreement was set up by one of Ferguson's export representatives, Peter Boyd-Brent, who also devised a paddy field cultivator that was made in small numbers by Balmer Lawrie. Few were sold as many of India's paddy fields had no subsoil and the tractors tended to sink, and the system proved too expensive to compete against the bullock and rake that was the usual method of cultivation.

The greatest number of Ferguson sales in India was in the north of the country, particularly in the Punjab, a large wheat-growing area that was dubbed 'the granary of India'. It has been claimed that Escorts sold 400 Ferguson tractors to the Punjab in one day. The tractors were also popular on the tea estates in Assam, where they were used with 3-ton trailers to transport the leaves from the plantations to the estate factory.

As normal-octane vaporising oil was freely available in India, most of the Ferguson tractors sold were TE-D20 models, although diesel tractors began to dominate sales in the later years. However, a number

RIGHT: *Harry Ferguson's North American plant at Detroit in Michigan. Known as Ferguson Park, it opened in 1948.*

LEFT:
The first Ferguson tractor to roll off the assembly line at Ferguson Park in Detroit on 11 October 1948.

of lamp-oil tractors were supplied through the Indian sales office for use in Ceylon.

The Ferguson tractor was evidently quite suited to the Indian climate and gave few problems. The only breakdowns were due to dust getting into the engine while the tractors were working with disc ploughs in the Punjab. This problem was simply overcome by cutting a hole in the tractor's hood, extending the air-intake pipe through it and fitting a domed pre-cleaner assembly. In extreme conditions, the cooling fan was also reversed to blow instead of suck to help keep the dust out of the radiator.

Bill Ketchell, a service manager from Coventry, took over from Roland Heath as executive director of Harry Ferguson (India) Ltd. in 1953. Ferguson accounted for around 80 per cent of the total tractor sales in India, and nearly 10,000 TE-20s had been sold in the country by 1956.

Following the formation of Massey Ferguson, Harry Nanda expressed an interest in manufacturing MF tractors under his Escort brand name. However, the negotiations fell through and the Nanda group began building an Escort tractor based on the Polish Ursus, later acquiring the licence to make Ford tractors in India. The licence to build Massey Ferguson tractors was eventually given to Tractor and Farm Equipment (TAFE) in 1960, which then established its manufacturing base in Madras.

Yugoslavia was another country with its own production facilities for Ferguson tractors. It was an association that began in 1955 after the Balkan country signed a licensing agreement on 16 October with both Massey-Harris-Ferguson and Frank Perkins Ltd. for the importation and eventual manufacture in Yugoslavia of tractors, implements, combines and engines. Many different makes of tractor had been evaluated in the country after the Second World War, but the Ferguson was deemed most suitable for Yugoslavia's needs.

Under the terms of the contract, 900 Ferguson tractors and 2,000 implements, including ploughs, disc harrows, subsoilers,

RIGHT: *Harry Ferguson addresses the workforce at the Detroit plant on the occasion of the launch of the American TO-20 tractor in October 1948.*

mowers and orchard equipment, were shipped to the country in 1955, followed by a further 2,800 tractors, 9,000 implements, 11,000 accessories and 100 Massey-Harris self-propelled combines during the following year. The total order was worth over £4.25 million. The tractors were exported in PKD form and were assembled by the Industrija Traktora I Masina (ITM) in Belgrade.

The agreement allowed for Yugoslavia to begin the progressive manufacture of Ferguson and Massey-Harris equipment under licence in its own factories, although no export was permitted. The tractors were built by ITM and fitted with Perkins engines, and were sold alongside Yugoslavian Zadrugar tractors from the Industrija Motora Rakovica (IMR) concern. Massey-Harris combines were made in Zemun by the Industrija Poljoprivrednih Masina under the Zmaj name.

Ferguson's representative in Yugoslavia was Bud White who liaised with ITM's general manager, Svetislav Milivojevic. Most of the Ferguson tractors were supplied to co-operative and state-owned farms. After the TE-20 was replaced by the FE-35 in 1956, Yugoslavia continued to build the new model under licence, but such was the demand for Ferguson equipment that the locally manufactured stock often had to be topped-up with machines imported from the UK. This saw another shipment, including 2,260 FE-35 tractors in CKD form, 12,500 implements and 240 combines, with a total value of £2.5 million leave Britain for the Balkans in May 1957. By 1959, Yugoslavia had amassed a 'population' of 10,000 Ferguson tractors in only four years. Incidentally, the ITM and IMR concerns later merged to form Industrija Motora I Traktora (IMT) which continued to manufacture Massey Ferguson tractors under licence until production was affected by the recent Balkan conflicts.

Harry Ferguson's most important overseas market was, of course, North America, but his interests in the USA had been jeopardised by Henry Ford ll's decision to halt the production

LEFT:
A 1948 Ferguson TO-20 tractor powered by a Continental Z-120 petrol engine. The American machine was virtually identical to the British TE-20 except for a few small detail changes. Note the different lugs to bolt the rear wheel rim to the centre.

of the Ford Ferguson tractor in 1947. Ferguson had a sales company, Harry Ferguson Incorporated, and an established dealer network, but alas no product to sell.

The staff at Harry Ferguson Incorporated had not been idle; the management under Horace D'Angelo was actively pursuing new production facilities, either through an association with another company (unsuccessful talks had already been held with both General Motors and Willys-Overland Motors) or by securing their own factory premises. The engineering group, run by Hermann Klem after John Chambers returned to the UK, was also evaluating new tractor designs.

One intriguing and little-known development to surface at this time was a mechanism to convert the Ferguson hydraulic system to position control. The details of the mechanism

LEFT:
A 1950 Ferguson TO-20 with the open or 'skeleton' steel wheels. The Ferguson badges on the hood sides were introduced during the first year of production.

were couched in a patent for a crane attachment that was filed by Harry Ferguson Incorporated on 25 November 1946, and assigned to Ernest Bunting, the hydraulic specialist in Hermann Klem's team.

The patent described the invention as 'a power device which can give an alternative or positional control, that is control in which the implement will be raised or lowered by the lift device to follow with fidelity the movement of a hand control lever'. The mechanism was no more than an external linkage with a special top link with a roller and cam lever operating as a fulcrum against the rear-axle casing to send signals back to the control valve. It was simple but crude, and among the many limitations of the system was the fact that the mechanism and linkage had to be removed and the normal top link replaced when changing back to draft control. Not surprisingly, the conversion was never marketed commercially.

In January 1948, Harry Ferguson Incorporated managed to acquire a suitable 72-acre site on Southfield Road in Detroit for a new tractor factory. As an interim measure, a consignment of around 15,000 Continental-powered TE-20 models was imported into the USA from Britain. Not only did this help to tide the dealers over and keep the Ferguson name in the market place until North American production could begin, it also allowed Harry Ferguson Ltd. in Coventry the pleasure of boasting that it had secured a $20 million order. In reality it was no more than a transaction between two arms of the same organisation, but it was still a welcome boost to British exports.

BELOW: *A sales brochure introducing the North American Ferguson TO-30 tractor that was launched in 1951.*

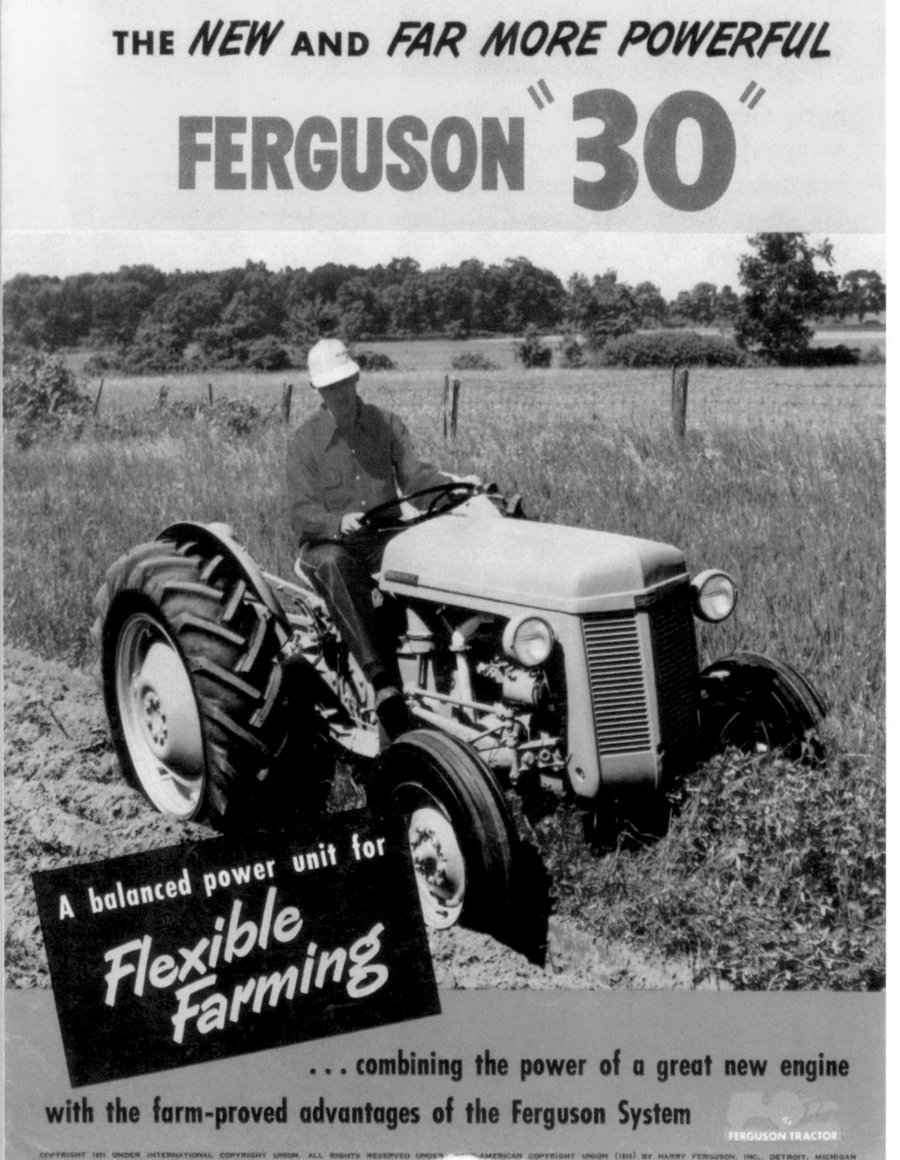

THE NEW AND FAR MORE POWERFUL
FERGUSON "30"
A balanced power unit for
Flexible Farming
... combining the power of a great new engine
with the farm-proved advantages of the Ferguson System
FERGUSON TRACTOR

Work on the Detroit site began on 13 February. The plant was named Ferguson Park and was miraculously completed by 26 July in time for Harry Ferguson to drive the first tractor off the assembly line on 11 October. The new American Ferguson was known as the TO-20, standing for Tractor Overseas. It was essentially a copy of the early British TE-20 and was fitted with the same Continental Z-120 petrol engine. The only real difference was that the TO-20 used components sourced in the USA, including Delco-Remy electrics and Donaldson or Vortox air-cleaners. The Borg-Warner Corporation was the main sub-contractor, supplying the Warner hydraulic pump and the Rockford clutch as well as building the gearbox and rear axle at its transmission plant in Detroit.

The job of organising the suppliers and co-ordinating the manufacture of the tractor fell to Albert Thornbrough, a graduate who had worked for the US Department of Agriculture in a research capacity during the war before joining Harry Ferguson Incorporated in 1946. He later became a director and vice-president of the company.

A first full-year production figure of 12,859 TO-20s during 1949 was a creditable start for the new American plant. This figure doubled

LEFT:
A 1951 Ferguson TO-30 – the successor to the TO-20 with a more powerful Continental Z-129 engine developing 29 hp.

LEFT:
The Ferguson TO-35 that was launched in 1955. Designed by Hermann Klemm, it had a six-speed gearbox, a revised hydraulic system and was powered by a Continental Z-134 petrol engine. The tractor is seen with a Ferguson high-speed rotary hoe.

ABOVE:
A Ferguson TO-35 featuring the later style decals as attempts were made to rationalise and integrate the Massey-Harris-Ferguson tractor ranges.

to 24,503 for 1950 and 33,517 tractors rolled out of the Detroit facility through 1951. However, the plant never reached the dizzy heights of Banner Lane production and sales were lower than expected.

A power increase with the adoption of the improved and larger Continental Z-129 engine that saw the Detroit Ferguson revamped as the TO-30 in August 1951 did little to alter matters. The American company remained a weak link within the Ferguson organisation as falling profits culminated in a loss for 1953, partly due to Harry Ferguson's insistence that selling prices were kept to a minimum, and partly because the tractors were competing against the very similar Ford models that were already well entrenched in the market place.

It is interesting to note that historically this was the only Ferguson tractor manufacturing facility to come directly under the control and

RIGHT:
The Ferguson 40, introduced in 1956, was a heavier and stronger version of the TO-35. It used the same engine and transmission, but had hub reductions on the rear axle and a beam-type front axle.

LEFT:
The Ferguson Hi-40 high clearance tractor. It is seen with the high-arch wide-front axle; a tricycle version was also available.

ownership of the Ferguson organisation. The other three plants worldwide, at Banner Lane and in India and France, were operating under the experienced guidance of the Standard Motor Company. It could be argued that Harry Ferguson's management team were more adept at marketing than at manufacturing, and that this was probably part of the problem in the USA. Whatever, the fact that his American interests were a financial disappointment is often cited as an important catalyst in Harry Ferguson's decision to sell out to Massey-Harris.

Other manufacturers, particularly Ford, were moving towards a policy of having one base design that could easily be modified from the standard agricultural tractor with a wide-front axle to a rowcrop, tricycle or utility machine, or fitted with different engines to provide a range of power classes. In an attempt to emulate this, Hermann Klemm introduced what has been dubbed as his 'meccano' concept, using one base design to develop a range of models. It was not a concept that would have pleased Harry Ferguson with his ideals of simplicity of design and construction, but it did appeal to the Massey-Harris hierarchy who were in the driving seat after the 1953 merger. In March 1954, Klemm was elevated to the position of chief engineer for Massey-Harris-Ferguson's North American products and was given the go-ahead to develop his concept further, starting with his TO-35 design that was a proposed replacement for the TO-30.

The Ferguson TO-35 tractor was launched in January 1955. At first glance, the new model looked little different to the old TO-30 except for a new livery. The sheet metalwork and wheels were still grey, but the engine and chassis were painted a dark metallic green colour. However, first appearances can be deceptive and under the skin much had been changed.

Power came from a Continental Z-134 petrol engine, a 133.8 cu in. (2,193 cc) unit developing 32 hp. The transmission had six forward and two reverse speeds. It incorporated an epicyclic reduction unit with a two-speed planetary gearset that was inserted in series with the output shaft of the main three-speed gearbox. The main gearbox had sliding spur-gears while helical-gears were used in the first reduction gear-train.

The hydraulic system was considerably altered from that of the TO-30 and much of its design was attributed to Ernest Bunting. The pump capacity and oil pressure were increased allowing faster operation and giving the tractor greater lift capacity at the lower links. The

ABOVE: *The Massey-Harris 50 was identical to the Ferguson 40 except that it was styled to suit the Massey-Harris dealers and was finished in a red and gold livery.*

system incorporated position control and was operated by two hand-levers in a dual quadrant, an arrangement that the company referred to as Quadramatic control. It was a much more sophisticated system than Bunting's earlier attempts at position control and relied on internal linkage.

A two-speed power take-off was fitted and both engine and ground speed could be selected through a sliding-gear clutch. A 'live' drive that allowed the power take-off and hydraulics to operate independently of the transmission was available. It was provided by a two-stage main clutch with the drive taken through a hollow extension of the gearbox countershaft. The brakes were improved with the pedals relocated to the right of the transmission, and a new recirculating ball-type steering mechanism, supplied by the Saginaw Steering Gear Division of General Motors, was fitted.

Two versions of the TO-35 were available; the standard model was priced at $2,263, while the de-luxe model, with 'live' power take-off and hydraulics, a cushioned tilting seat and a tractormeter, cost an extra $100. Power steering, optional tyre equipment and power-adjustable rear wheels were also available.

As many of the North American Massey-Harris dealers had asked if they could sell a tractor incorporating the Ferguson hydraulic system, Klemm put his 'meccano' concept to work and produced a heavier and stronger version of the TO-35 known as the Massey-Harris 50. It had the same engine, transmission and hydraulic system, called Hydramic Power on the Massey, as the TO-35, but was built to a rowcrop configuration with a beam-type front axle. It was also 4 in. longer and had hub reductions on the rear axle.

The red and gold Massey-Harris 50 was unveiled in May 1956 along with the F-40 model, a similar version of the same tractor with different sheet metal for the Ferguson dealers. Although the Ferguson 40 was launched in the same grey and green livery as the TO-35, this was soon changed to a grey and cream colour scheme with red grille inserts. The company's sales literature ambitiously called it 'bold and beautiful'.

The standard Ferguson 40 cost $2,351. There were also three Hi-40 high-clearance models available with a high-arch wide-front axle or in a tricycle arrangement with dual or single front wheels. The F-40 was the last North American tractor to be identified as a Ferguson. It was dropped from the line within a year while the Massey-Harris 50 continued in production as the Massey Ferguson 50.

Other Ferguson models were planned, and would have gone ahead had not the decision been made to integrate the Ferguson and Massey-Harris lines. A more powerful version of the F-40, fitted with a Continental G-176 engine, would have become the Ferguson 65 while the Massey-Harris version of the same tractor would have been sold as the 75 model. Heavier Ferguson 80 and the identical Massey-Harris 90 models with a 242 cu in. Continental engine were also in the pipeline in 1956.

The merging of the product lines and the creation of Massey Ferguson in December 1957 killed off the twin model policy. The 65/75 model became the Massey Ferguson 65 with Massey-Harris styling for North America, while the 80/90 design was launched as the Massey Ferguson 85. It must be noted that the American Ferguson tractors were complemented by an equipment line equal to, if not greater than, that offered in the UK, including a unique side-mounted baler and forage harvester.

BELOW:
Amid the confusion during the amalgamation and rationalisation of the Ferguson and Massey-Harris product lines, some North American machines were provisionally badged as Massey-Harris-Ferguson tractors, but it is not thought that they were ever marketed as such. From left to right are the 50 and the new 65 and 85 models, all probably pre-production tractors with Continental petrol engines.

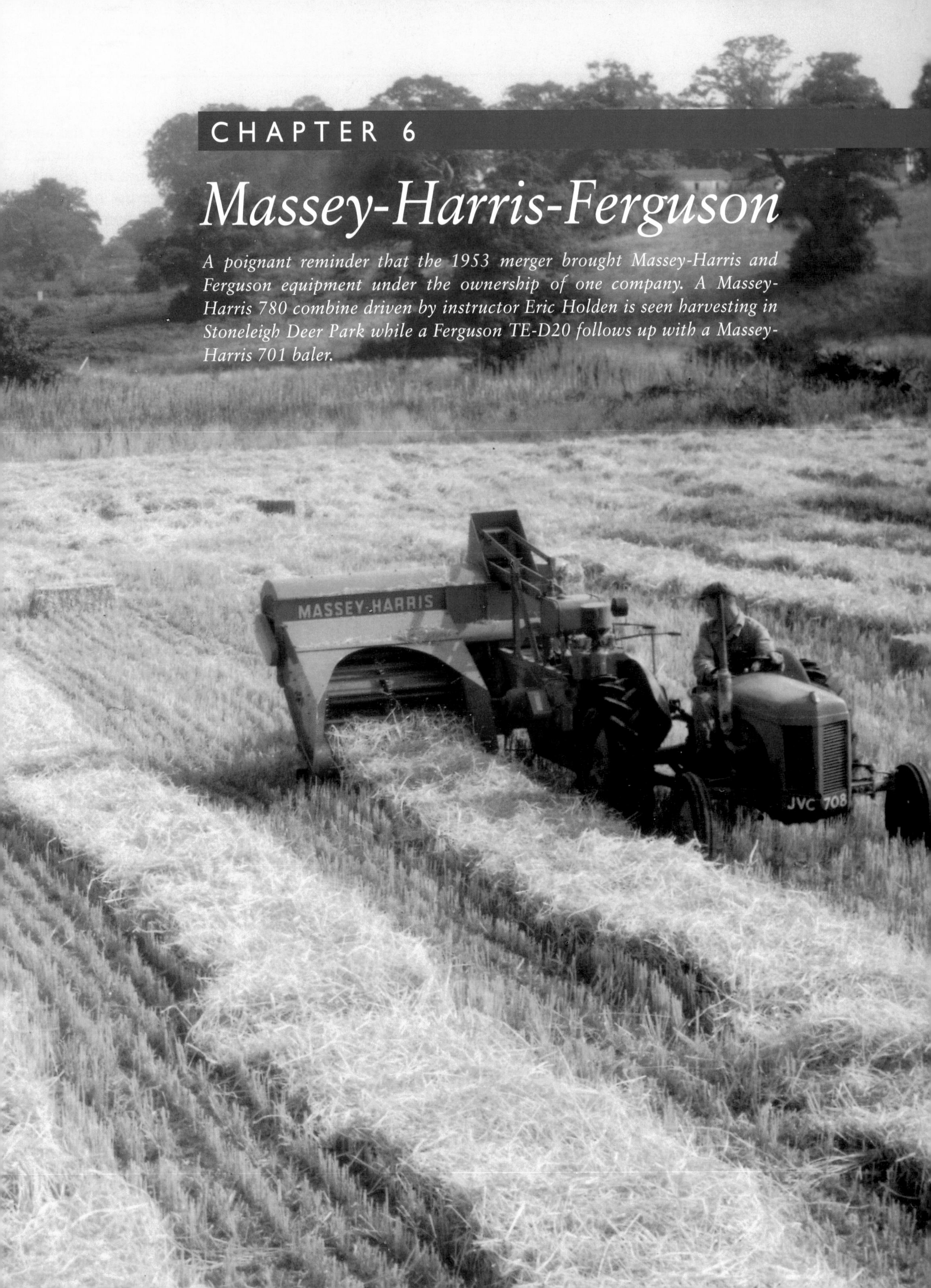

CHAPTER 6

Massey-Harris-Ferguson

A poignant reminder that the 1953 merger brought Massey-Harris and Ferguson equipment under the ownership of one company. A Massey-Harris 780 combine driven by instructor Eric Holden is seen harvesting in Stoneleigh Deer Park while a Ferguson TE-D20 follows up with a Massey-Harris 701 baler.

ABOVE: *The 1953 merger brought Massey-Harris and Ferguson equipment under the umbrella of the new Massey-Harris-Ferguson organisation. The Massey-Harris tractor in production at the time of the merger was the 744D model. Built at Kilmarnock, it was powered by a Perkins P6 diesel engine.*

Massey-Harris was an old established firm that had been in the farm machinery business for over 100 years. It was formed from the amalgamation of two Canadian firms founded by Daniel Massey and Alanson Harris, both pioneers in agricultural engineering. Like Ferguson, the company had a full product line, but was particularly recognised for its combine harvesters, balers and hay harvesting equipment. The firm had a strong presence in Britain and had opened a factory at Kilmarnock in Scotland in 1949.

Initially, the talks that led up to the merger centred on Massey-Harris agreeing to manufacture a proposed Ferguson combine at Kilmarnock and getting the use of Standard's TE-20 engines to power its equipment in return. However, it seems that both sides already had a hidden agenda; Ferguson was keen to amalgamate with a global corporation, and Massey-Harris had been under pressure from its dealers, particularly in the UK, to improve its tractor range and tillage line. The Canadian firm had already made several unsuccessful attempts at reaching a marketing agreement with other tractor manufacturers, including Nuffield, David Brown and Hanomag.

The Massey-Harris 744D tractor that was currently in production in Britain was something of a compromise, and was no more than a North American 44 chassis fitted with a Perkins P6 diesel engine. An improved version, the 745 model, designed by Pat Mulholland and fitted with a Perkins L4 engine, was in the pipeline. However, the British Massey-Harris tractors, built at Kilmarnock using transmissions supplied by Beans Industries in the Midlands, never really achieved outstanding sales with the exception of one major export deal to the Sudan.

Bill Nunn of Frank J. Nunn & Sons Ltd., one of the leading Massey-Harris dealers in East Anglia, remembers writing to suggest that if the company wanted them to stay exclusively

RIGHT: *Although the dealer franchises of Ferguson and Massey-Harris still remained separate after the merger, the new corporate name, Massey-Harris-Ferguson, appeared on the company's own vehicles, such as this Vanguard van used by the field-test staff. Seen with the van in 1954 are Colin Steventon, Charlie Nicholson and Roger Reid.*

LEFT:
Following the merger, the activities of the training centre at Stoneleigh were extended to offer courses to Massey-Harris dealer personnel. One of the instructors, Ted Dowdeswell, in the white overalls, shows two overseas students how to adjust a disc plough. The two Massey-Harris tractors are a North American 55K model (on the left), and the new British 745 that was launched in 1954.

Massey agents, then it should provide them with a better tractor and consider buying out Harry Ferguson. He never dreamt that this would actually happen, and still wonders if his letter was a catalyst in the chain of events that led to the formation of Massey-Harris-Ferguson.

Harry Ferguson's offer to sell his company was made to Massey-Harris's president, James Duncan, at a meeting held at Abbotswood on 4 August 1953. Duncan, the son of a Scottish agricultural engineer who had moved to France in 1887 to become an agent for the Massey Manufacturing Company, had been president of the organisation since 1941. He had joined Massey-Harris in 1910 and had held several senior appointments in various European branches before moving to the head offices in Toronto where he was made general manager in 1935. Both Ferguson and Duncan had spent a lifetime in agricultural engineering, and this mutual bond paved the way to an agreement.

After the terms of the deal were hammered out, it was agreed that Harry Ferguson would receive $16 million worth of Massey-Harris shares in exchange for the operating Ferguson companies. It appears that when several of the

LEFT:
The 'big Fergie' - a prototype LTX tractor fitted with the diesel engine on test in 1952. The field-test team's Vanguard van can be seen in the background. The location is possibly Wood Farm at Ufton in Warwickshire.

RIGHT:
A four-cylinder LTX diesel engine on one of the test-beds at Fletchampstead. It is connected to an electric dynamometer that accurately measured the torque so that its brake horsepower could be calculated. The diesel was rated at 60 bhp.

Massey-Harris directors went to close the deal with Ferguson, they suddenly realised that their French interests had been accidentally left out of the valuation and that Harry was due another $1 million. Rather than see the deal fall through, Ferguson suggested tossing a coin over the difference and lost.

A joint press statement announcing the merger was issued on 17 August.

The new organisation was to be known as Massey-Harris-Ferguson Ltd. with Duncan as president. Harry Ferguson was offered the position of chairman of the board of the parent company while still retaining the responsibility for the design, engineering and application of the Ferguson System.

The new joint company effectively came into being on 30 January 1954. An eastern hemisphere division, with a former Ferguson man, Eric Young, as its managing director, was created as an umbrella body under which the UK subsidiary companies were grouped. Massey-Harris Ltd. became Massey-Harris-Ferguson (Manufacturing) Ltd., while Harry Ferguson Ltd. changed its name to Massey-Harris-Ferguson (Sales) Ltd. with Trevor Knox as its sales director. Massey-Harris-Ferguson (Engineering) Ltd. was responsible for the research and development of Ferguson products. John Chambers was the chief engineer and director of this company assisted by two other County Down men, Alex Patterson, who was superintendent of engineering, and James

RIGHT:
The final styling mock-up of the LTX in the engineering workshop at Fletchampstead. The tractor was badged as the Ferguson TE-60. The back-end of a prototype Ferguson combine can just be seen on the left.

LEFT:
The Ferguson TE-60 is compared to the Massey-Harris 50. Note the other LTX prototypes in the background.

McNiece, the chief draughtsman. Alan Botwood and Horace D'Angelo had disagreed with the merger and had left the company soon afterwards.

Naturally, the merger led to some pooling of resources, and the activities of the training centre at Stoneleigh were also extended to include courses for Massey-Harris personnel. Integrating the training facilities was no mean task, and it was organised by Captain Duncan Hill who had taken over from Dick Chambers as manager of the school in 1951. Hill was an ex-naval man who had a reputation for running a very tight ship.

Rationalisation of manufacturing facilities saw some badge-engineering take place with Massey-Harris equipment painted grey for the Ferguson dealers and vice-versa. However, the Massey-Harris and Ferguson brand names were retained with the product lines and franchises kept strictly separate. Most dealers remained exclusively Massey-Harris or Ferguson agents, and were not allowed to use the Massey-Harris-Ferguson nomenclature or trademark in any shape or form, as they were reminded by a stern bulletin circulated on 17 March 1954.

There was some uneasiness at the Standard Motor Company as to how the merger would affect its arrangement to manufacture Ferguson tractors. The original agreement with Harry Ferguson had only just over two years to run, and the company was concerned that Massey-Harris would move tractor production to another of its factories, such as Kilmarnock, after the current contract expired. It was an unlikely scenario as no other MHF plant had

LEFT:
This TE-F20 model was the half-millionth Ferguson tractor to be produced at Banner Lane. It was built in March 1956 and is seen outside the plant with Eric Young, the managing director of Massey-Harris-Ferguson's eastern hemisphere division, in the driving seat for the occasion. Next to him is Alick Dick, the Standard Motor Company's managing director.

RIGHT: *The second Ferguson tractor to arrive in the Antarctic, this TE-A20, fitted with Bombardier half-tracks and a high-lift loader, was used at the ANARE station on Heard Island and is seen unloading supplies from a DUKW amphibious vehicle in 1954.*

the spare capacity to cope with Ferguson production, but just in case Banner Lane was left empty, Standard took tentative steps towards planning its own line of tractors.

Fears were allayed after Sir John Black and James Duncan reached a new agreement for Standard to continue Ferguson production for another twelve years. A joint statement issued at the end of 1953 detailed the success of the original agreement between the Standard Motor Company and Harry Ferguson Ltd. that saw the two companies co-operate to build and sell over 350,000 tractors. It went on to say that current production was running at a rate of around 60,000 a year, and that both Standard and Massey-Harris-Ferguson wanted 'to ensure the continuance of this co-operation in the manufacture and sale of Ferguson tractors throughout the world, for their mutual interest.'

The agreement covered not just UK production but manufacturing arrangements with Standard subsidiaries in France and India as well. The statement also said that Standard was to 'undertake the manufacture of a new and larger tractor'. This was the second time that a 'big Fergie' had been publicly hinted at over the previous few months.

The British trade journal, *Farm Implement and Machinery Review*, reporting the MHF merger in its September 1953 issue, used the headline 'Secret revolutionary machines will be marketed by the new Massey-Harris-Ferguson amalgamation'. This followed a statement issued by Harry Ferguson giving his side of the discussions that led up to the merger, saying that he disclosed to James Duncan 'in full all the new

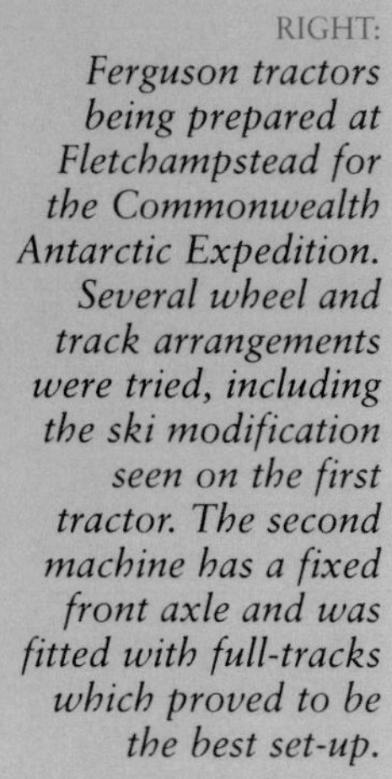

RIGHT: *Ferguson tractors being prepared at Fletchampstead for the Commonwealth Antarctic Expedition. Several wheel and track arrangements were tried, including the ski modification seen on the first tractor. The second machine has a fixed front axle and was fitted with full-tracks which proved to be the best set-up.*

inventions, including the new larger tractor (which had been so secretly developed in the past ten years) and a new small combined-harvester of revolutionary design (that we have started to develop)'. The combine we shall return to later, but the proposed new tractor was to play a pivotal role in the relationship between Harry Ferguson and the MHF board.

In truth, the new large tractor had not been under development for over ten years as Harry Ferguson suggested. However, he would probably argue that the original concept was born out of the experimental 4P project that saw the prototype larger Ford Ferguson tractor appear in 1944. The 4P was separately evaluated by Ford and Ferguson, both of which believed the tractor was unsuitable for production and shelved the design. However, it was not the last that we would see of a 'big Fergie'.

As discussed earlier, the size of the TE-20 was limited by the design of its back axle. However, the post-war advances in farm mechanisation

ABOVE: *Ferguson tractors equipped with full-tracks moving supplies for the Commonwealth Antarctic Expedition at Scott Base on the Ross Sea coast of Antarctica in 1956.*

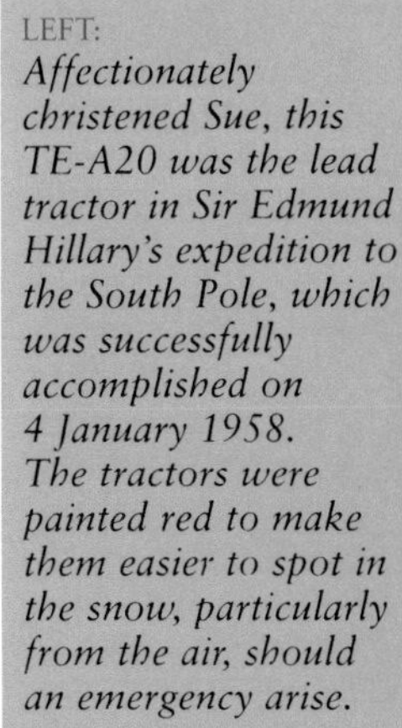

LEFT: *Affectionately christened Sue, this TE-A20 was the lead tractor in Sir Edmund Hillary's expedition to the South Pole, which was successfully accomplished on 4 January 1958. The tractors were painted red to make them easier to spot in the snow, particularly from the air, should an emergency arise.*

RIGHT:
Sue was eventually returned to Coventry in 1965 and is now on display in the Massey Ferguson museum at Banner Lane. The tractor has recently been restored, which in many ways seems a shame as it has lost much of its original character.

BELOW:
Following the success of the flexible tracks in Antarctica, the half-track was manufactured under licence in Britain and offered as genuine Ferguson equipment from 1956.

meant that the Ferguson tractor was in danger of becoming outclassed. Other manufacturers were building larger and more powerful tractors and both the Ferguson engineering staff and salespeople knew that they had to do something to stay competitive. It was a view that even Harry Ferguson had to reluctantly endorse, giving his engineers the go-ahead to develop a larger tractor in 1948.

The engineering team for the 'big Fergie' project, codenamed LTX for 'large tractor experimental', was led by John Chambers and Alex Senkowski. Their design brief was simple: produce a more powerful tractor that could handle larger implements while still embodying all the main principles of the TE-20.

Planning the LTX began in September 1948. The gearbox was a similar but stronger version of that used on the TE-20 with constant-mesh helical-gears, but had an extra creeper gear making five forward speeds and one reverse. A two-stage clutch provided 'live' drive to the power take-off and hydraulics. Senkowski also radically redesigned the hydraulic system which was powered by a three-cylinder pump and had a lift capacity of 6,000 lb (2,720 kg). A lot of thought was given to the strength of the rear axle, which featured double-reduction gearing

LEFT:
The Ferguson side-delivery rake that was one of several new implements introduced during 1954.

with inboard bull-gears. A two-speed power take-off and differential lock completed the specification.

The development of the engines for the LTX tractor was nearly a separate project of its own. The idea was to evolve a line of two-, four- and six-cylinder power units from the one basic engine design to simplify production and allow a degree of commonality of parts. The engines were designed 'in-house' and the engineer responsible was Bill Harrow, whom Senkowski had poached from another Coventry firm, Transport Vehicles (Daimler) Ltd., where he had been involved in the development of a new diesel engine for double-decker buses.

The six-cylinder, provisionally planned as a 100 hp unit, existed only on paper. One example of the twin-cylinder engine was built as a direct-injection diesel, but did not appear until the period of the MHF merger. It evidently performed well and was mooted as a possible replacement for the petrol engine fitted to the Massey-Harris Pony tractor, possibly for the French market. However, the decision was made not to use it, and the French Pony, the 820 model, was powered by a Hanomag diesel.

The only engine to appear in any quantity was the four-cylinder. It was designed to power

LEFT:
Ferguson equipment being loaded at Coventry station for shipment by rail to London for the 1956 Smithfield Show. The TE-F20 nearest the camera is fitted with the idler wheels for the half-track attachment while the other tractor has the low-volume sprayer.

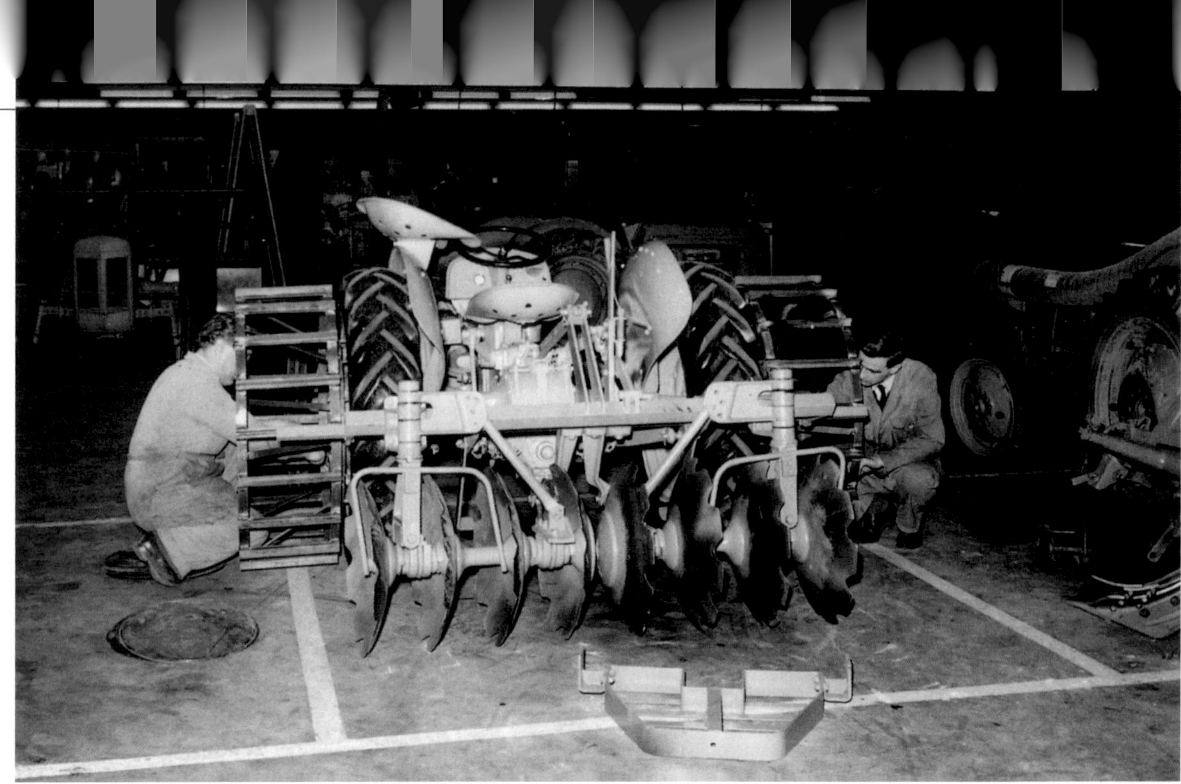

RIGHT:
Prototype cage wheels being fitted to a Ferguson tractor with a reversible heavy-duty disc harrow in the engineering workshop at Fletchampstead. A similar arrangement was used for the wet paddy cultivation system introduced in 1954. The spare seat on the tractor's fender was for Harry Ferguson to ride on.

BELOW:
The first prototype Ferguson combine on test during the 1954 harvest. Note that it is mounted on the right-hand side of the tractor. Seen with the combine (left to right) are Witol Czarnocki, George Belkowski, fitters Bill Halford and Joe Curzon. Geoff Metcalfe is in the tractor seat, while field-test engineer Charles Nicholson is behind him. Dick Dowdeswell is sitting on top of the combine, and the lady on the back is Michael Stoltenberg-Blom's wife, Jean.

the LTX tractor and both petrol and diesel versions were built. The petrol engine was a 3,113 cc (190 cu in.) unit and was fitted with a Zenith carburettor. The diesel had an increased capacity of 3,671 cc (224 cu in.) and used a CAV type-BPE fuel injection pump similar to that fitted to the Standard 20C engine.

Beans Industries Ltd. of Tipton in Staffordshire supplied the block castings and assembled the first batch of twenty engines, which were then sent to Fletchampstead for testing. Beans were another Standard Motor Company subsidiary and had built cars and commercial vehicles as a separate concern during the 1920s and 1930s. However, they now were better known as accessory and component suppliers than engine manufacturers, and the test programme reportedly uncovered a small number of minor faults in the first batch of engines that had to be rectified by Ferguson's fitters who stripped and rebuilt several of the power units.

Harry Ferguson drove the first prototype LTX, a petrol tractor, out to work with a three-furrow plough in April 1949. Three further petrol tractors followed and were later joined by two diesel LTX models. The prototypes were all built in the engineering department at Fletchampstead by a team led by Alex Patterson. The fitters working on the LTX included Nibby

Newbold, Frankie Ball and Len Goodenough.

Like all Ferguson equipment, the LTX tractors were subjected to the most rigorous test programme. The testing was carried out on blue clay at Wood Farm, Ufton, near Leamington Spa in Warwickshire. Each tractor was running on test from 6 am in the morning to 10 pm at night daily for three years. The field-test engineers worked an eight-hour shift each, changing over at 2 pm in the afternoon. Jack Bibby, Dick Dowdeswell, Nigel Liney and Colin Steventon all took a turn at the wheel. Most of the day was spent continuously ploughing with a three-furrow plough, while the tractor was used for tree pulling at night. It was tiring work; there was no power steering, it was a heavy tractor and there was a lot of kick-back from the steering wheel.

The field-test engineers tried to break the LTX and failed. They only managed to pull the lugs off several sets of tyres. There are not enough superlatives to describe their opinion of the tractor. The machine was simply phenomenal. The petrol engine was good, but the torque of the diesel engine was unbelievable, and the power it developed seemed far in excess of its rated 60 hp. The combination of the engine, hydraulics, weight and balance of the tractor made it an unbeatable package. The LTX would handle five-furrows with ease and reportedly scared the pants off Ford's engineers who were rumoured to be hiding behind the hedge with their binoculars every time the 'big Fergie' ventured out into a field.

LEFT: *A rare photograph of the first prototype Ferguson combine working near Evesham in 1954. The driver is Bill Halford.*

By 1953, Harry Ferguson felt that the LTX was ready for production. A batch of fifty engines had been completed, the final styling mock-ups of the proposed sheet metalwork were prepared and the tractor was provisionally badged as the Ferguson TE-60. However, with demand for the TE-20 still very high, and Banner Lane running at full capacity, the new tractor's introduction had to be delayed. The Ferguson dealers were informed about the new model in May and were given possible delivery dates for spring 1954. During the negotiations

BELOW: *Later versions of the Ferguson combine were all mounted on the left of the tractor. This is believed to be the third prototype harvesting in Australia, and is considerably different from several of the versions that followed. Note the spark-arrester fitted to the top of the tractor's exhaust pipe.*

ABOVE:
The fourth prototype Ferguson combine at Fletchampstead. The tractor is fitted with an epicyclic reduction box that can be seen sandwiched between the gearbox and rear transmission. Note the complicated chain drive to the pick-up reel.

surrounding the merger in August, Harry Ferguson arranged for the Massey-Harris board to see the LTX in action. The demonstration was held at Barford near Warwick. The 'big Fergie' with Jack Bibby at the wheel was pitted against Massey-Harris 744 and prototype 745 tractors, which it easily outperformed and outclassed.

After the merger, Harry Ferguson took it for granted that the LTX would go into production, but the MHF board remained unconvinced of the large tractor's merit and demanded certain changes. They felt the design was unsuitable for the North American market, particularly as there was no provision for a tricycle version. Duncan also wanted to use Perkins or Continental power units in preference to the unproven LTX engines. Typically, Harry Ferguson refused to alter or compromise the 'big Fergie' and the decision on its future was deferred to be eventually decided at a product planning meeting that took place in San Antonio, Texas, during the second week of March 1954. It was left to Hermann Klemm to have the final say on the LTX, and he naturally

RIGHT:
The fifth prototype of the Ferguson combine working near Stratford-on-Avon in 1955. The driver was Colin Wright, a fitter from the engineering department. The combine had a 7½ ft cut, and a Sturmey-Archer gear change was incorporated in the large sprocket nearest the tractor to give eighteen different reel speeds.

ABOVE:
A Ferguson combine, possibly the No.5 prototype, leaving the Fletchampstead works en route to Scotland for field trials.

ruled in favour of his TO-35 'meccano' concept for future models.

Even today, the scrapping of the LTX still rankles among many former Ferguson employees. They believe the decision to shelve the project was made for political rather than engineering reasons, and cite rivalry within the company between North America and the UK, or red against grey, as the real reason. Whatever, if their opinion of the tractor is to be believed, it looks as if it would have blown all the opposition out of the water had it gone into production.

The dropping of the LTX project was also the final straw in a deteriorating relationship between Harry Ferguson and the MHF board. He had already clashed with Duncan over pricing structures and, saddened by the train of events, he tendered his resignation as chairman of Massey-Harris-Ferguson Ltd. with effect from 7 July 1954 to pursue other interests in the motor industry. It has been suggested, however, that Ferguson was ready to break from MHF and stage-managed the whole LTX dispute to get the company to buy out his shares to finance other interests.

The loss of its figurehead did little to affect the MHF ship that still sailed on regardless. Duncan took over the vacated chairmanship, and Albert Thornbrough was appointed to oversee and co-ordinate the Ferguson tractor programmes in the USA, UK and France. Thornbrough was one ex-Ferguson employee who was moving fast up the MHF corporate ladder and was promoted to executive vice-president of the organisation in 1955.

The change of ownership of the company had done nothing to diminish the popularity of Ferguson equipment. By 1954, it was estimated that 825,000 Ferguson tractors and over 2.5 million implements were in use worldwide. More TE-20 tractors were sold in the March of that year than in any other month since production began at Banner Lane. Home sales had been their highest for over a year, but the demand from the export markets was such that many orders could not be met and were held over into April. Even so, the March production figure for the Coventry plant was 7,164 tractors - a figure that helped contribute to the British Ferguson passing the half-million milestone two years later. The 500,000th TE-20 was driven off

RIGHT:
The sixth prototype on trial in 1955 at Major Harvey-Bathouse's Castle Farm near Ledbury in Herefordshire. Colin Wright is driving and Dick Dowdeswell is bagging-off on the back. Dust was a problem for the driver who also had to cope with poor visibility and the lack of power steering.

the Banner Lane assembly line by Alick Dick who had taken over as managing director of the Standard Motor Company after Sir John Black unexpectedly resigned in January 1954.

At the time, the public reason for Sir John's retirement was given as ill health; he was still recovering from serious injuries received the previous autumn when he was involved in a road accident as a passenger in a Swallow Doretti sports car. The Doretti, which used Standard-Triumph running gear, had been on a demonstration run when it hit another car outside the Banner Lane works.

Unofficially, Black was ousted in a boardroom power struggle; he had become increasingly autocratic in his style of management and had even threatened to sack Ted Grinham. The final straw came when he signed the twelve-year deal to continue Ferguson tractor production without consulting the Standard board who then called for his resignation. Alick Dick, who had joined Standard as an apprentice in 1934 and worked his way up to become Sir John's deputy, was his natural successor, while Grinham was appointed as the new deputy managing director.

TE-20 tractors had been sent to most corners of the globe, but none were further from home or working in more adverse conditions than those operating in the sub-zero temperatures of

RIGHT:
A Prototype Ferguson combine No.7, modified into a tanker model and fitted to one of three American TO-35 tractors sent to the UK for evaluation in 1956. Seen working in the Cotswolds, it is believed that this prototype was later sent to the USA for trials.

the Antarctic. The first Ferguson tractor to set wheel on the world's coldest and most inaccessible continent arrived on 6 January 1954 at Mawson station as a support vehicle to the Australian National Antarctic Research Expedition (ANARE). It was modified to work at temperatures as low as minus 50 degrees Centigrade and was fitted with an insulated cab and Canadian Bombardier half-tracks to cope with the snow and ice. Several items of Ferguson equipment went with the tractor, including a winch and a high-lift loader for moving and unloading stores from the supply ship, and an earth scoop and blade-terracer for clearing snow and preparing landing strips. A second similarly equipped tractor was also supplied for use at the ANARE meteorological station on Heard Island.

The success of these two little 'Fergies' in coping with the polar conditions led to the TE-20 being chosen later as the ideal machine for setting up base camps and supply dumps for the Commonwealth Antarctic Expedition. Five tractors were supplied to Scott Base, and three of these were used by a team under the famous climber and explorer, Sir Edmund Hillary, to establish a supply chain of stores for a crossing of the continent from the Weddell Sea by a British expedition led by Vivian Fuchs. By the autumn of 1956, there were twelve Ferguson tractors working at various bases across the polar continent.

Many of the tractors were prepared in the engineering department at Fletchampstead prior to the expedition. The tractors, all TE-A20 petrol models, were fitted with heavy-duty batteries and starter motors, and the electrical wiring was insulated against the cold. The engine sump and transmission were filled with a low-viscosity (SAE 5W grade) refined mineral oil designed for extreme temperatures while the radiator coolant was 50 per cent ethylene-glycol antifreeze. As a further measure to keep the cooling system from freezing, the fan was reversed to blow warm air from the engine through the radiator instead of sucking in cold air.

Trials with the Antarctic tractors showed that the action of the Bombardier half-tracks tended to exert a downward force on the front axle, causing the front wheels to sink in soft snow. Several modifications were suggested to overcome this, including replacing the TE-20's front wheels with a pair of skis. The most successful arrangement involved extending the tracks around the front wheels, with the steering locked in the straight-ahead position, and then

LEFT:
This is thought to be the eighth and probably last prototype Ferguson combine, and the one that was used for trials in Europe. It had a revised and simplified chain drive to the pick-up reel. Note that the tractor's hood has been removed to help keep the engine cool while harvesting.

using the independent brakes to steer the tractor. When Sir Edmund Hillary, who the Ferguson engineers soon discovered was not one to mince words, was told that the modification could make the tractors difficult to steer, he evidently retorted, 'What do we want to steer the (expletive) things for, we're going straight all the way to the (expletive) South Pole.'

And it was straight to the South Pole that the Ferguson tractors went. After a year's work at the base camps, and having successfully established the supply chain, Hillary decided to push his three TE-20s on to the Pole. The tractors, painted red to make them visible in the snow, were roped together for safety and travelled in line for the journey. They set out on 14 October 1957 and arrived at the South Pole on 4 January 1958. It was a remarkable achievement for both men and machines. The tractors had covered over 1,200 miles of some of the most inhospitable terrain on earth, coping with

ABOVE: *The Ferguson rowcrop thinner for mechanically gapping sugar beet plants on trial with Colin Steventon at the wheel.*

RIGHT: *The Ferguson prototype sugar beet harvester was a two-stage system. The first unit consisted of a tractor fitted with a side-mounted topper and a rear-mounted top-saver that gathered up the beet tops and elevated them into a cart.*

LEFT:
The second stage of the Ferguson sugar beet harvester was this mounted lifter-loader unit seen on trial at Stoneleigh. It worked well in good conditions but performed badly in the wet.

snow, ice, blizzards and extreme cold while climbing over 10,000 ft to reach the polar plateau, all without any major mechanical problems.

The Bombardier flexible half-track conversion as used in Antarctica was also available in the UK for a little over £100. Imported from Canada, it consisted of reinforced rubber belts joined by steel cross-members. The belts were fitted around the Ferguson's rear tyres and tensioned by a pair of idler wheels in front of the rear wheels. This conversion was later manufactured in Britain and offered as genuine Ferguson equipment from 1956.

The range of Ferguson implements, accessories and equipment still expanded under Massey-Harris-Ferguson's ownership. The years of 1953 and 1954 saw the introduction of the rear-mounted crane, the Hydrovane compressor, available with a hedge-cutter attachment supplied by R. M. Marples, the Electromatic hammer-mill and the side-delivery rake. The Cutrake, invented by a farmer from Stratford-on-Avon for cutting, loading and carting kale, appeared in 1955 and was followed by the spinner broadcaster and a fork-lift attachment the next year. The low-volume sprayer was joined by a medium-pressure sprayer, and in 1957 Massey-Harris-Ferguson took over the manufacture and sale of Fisons Pest Control's entire range of crop spraying machines.

The export markets were catered for by the introduction of the disc-terracer for constructing irrigation ditches and water channels, and the polydisc seeder and combined cultivator for seedbed preparation in tropical countries, both launched in 1954. The following year, the company introduced its system of wet paddy cutivation that had been pioneered in India by Peter Boyd-Brent. It consisted of an

BELOW:
A second version of the top-saver with an improved belt drive to the elevator. The photograph was taken in the test field at the rear of the Fletchampstead premises. Standard's engine works at the Canley plant can be seen on the left in the background.

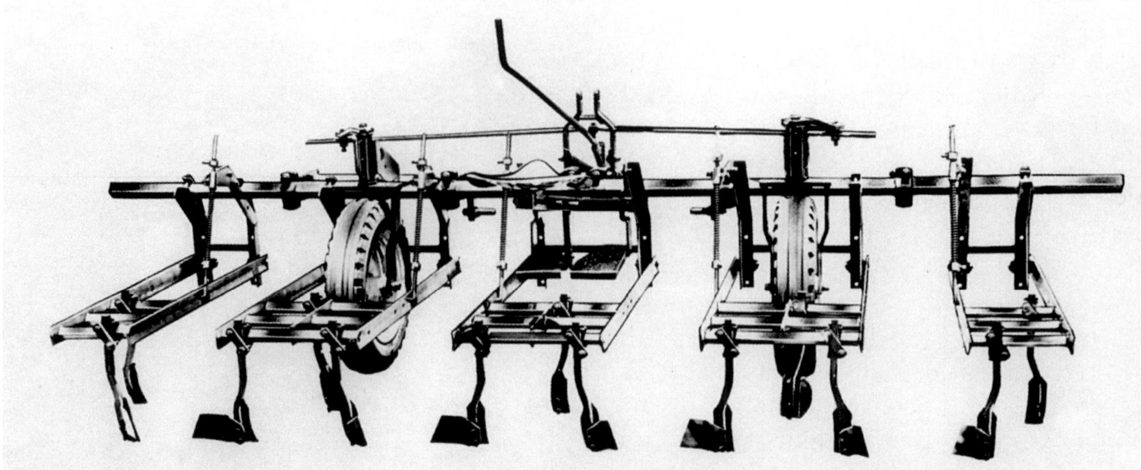

RIGHT:
A steerage hoe for maize - one of a small number of Ferguson implements that were developed but not put into production.

adapted mounted tandem disc harrow with levelling boards for 'puddling' or cultivating flooded rice fields. The tractor had to be modified to cope with the wet paddy fields, and was fitted with cage wheels, an upright exhaust, special waterproof brake shoes and brass plugs to seal the transmission housing. An area of ground at Stoneleigh was flooded and turned into a paddy field so that the system could be tested.

Without a doubt, the most fascinating piece of Ferguson equipment to appear during this period was the combine harvester that Harry Ferguson referred to in his statement following the 1953 merger. Ferguson's master plan had been to offer a machine to suit every farming need, and of course this had to include harvesting equipment. The combine was the brainchild of one of the training instructors, Michael Stoltenberg-Blom, who sketched the original design out in his notebook. He showed the sketch to Harry Ferguson who simply said 'Build it!'

The combine was designed to be a 'wrap-around' machine of the type later associated with JF of Denmark. It was mounted directly on to a TE-20 tractor that acted as the power unit. The idea was to provide an inexpensive self-propelled combine that had fewer working parts and was simple to maintain and repair.

BELOW:
A prototype Ferguson FE-35 tractor with the sugar beet lifter. The tractor, a petrol model with the new 87 mm engine, is seen in the yard at Fletchampstead in front of the Standard Motor Company's sportsground.

There was also the added advantage that the tractor could be released for other duties outside of the harvesting season.

George Belkowski was put in charge of the project as the combine development engineer while Geoff Metcalfe headed the design team. The first prototype combine, completed in 1953, was mounted on the right of the tractor. It was used during the 1954 harvest, but the initial field trials proved disappointing. The attachment was far too heavy for the tractor, and one TE-20 broke its half-shafts while travelling with the combine to a farm for trials. Repairs were made and the half-shafts were replaced, but the combine then promptly threw all of its straw walkers out of the back once harvesting began.

The later Ferguson combines were all much lighter and were mounted on the left of the tractor. Stan Hockey took over the design team and advice was sought from Massey-Harris's combine design office in Kilmarnock, and in particular, from the development engineer, Les Pierce. As a result of this input, the Ferguson prototypes became more compact and incorporated some Massey-Harris layouts.

The combine attachment weighed 2,300 lb and was supported on two carrier-brackets. One bracket was mounted on to the tractor's front axle while the second was in the form of a cradle fixed to an extension of the left-hand rear axle. The TE-20 drove the combine mechanism by vee-belts taken off the belt pulley attachment, and was fitted with an epicyclic reduction box to give more suitable forward speeds for harvesting. Incidentally, the epicyclic box was offered as a Ferguson accessory from 1955.

The Ferguson combine's cutting and threshing mechanisms were very similar to those used on the Massey-Harris 735; it had three straw walkers but its cut was wider at 7½ ft. Unique features of the design included a triple-beater system to feed the crop to the drum and a re-thresher mechanism to return the tailings to the grain pan. An eighteen-speed pick-up reel drive was provided by a Sturmey-Archer gear change, a heavy-duty version of that fitted to racing bicycles. This gear mechanism proved to be over-complicated and unreliable and was dropped from some of the later prototype models.

ABOVE:
Fitter Nibby Newbold working on an FE-35 tractor at the Fletchampstead engineering works.

A total of eight prototype Ferguson combines were made over three years and both bagger and tanker versions were built. They were trialled across the UK at farms in areas as far apart as Herefordshire and Scotland, and were sent to France, Germany, Holland, Denmark, Sweden, the USA and Australia. The machines evidently performed well, but there were several drawbacks to the design, not least the restricted visibility that meant the driver had to stand up to see the combine bed. At the end of the day, leg muscles would be nearly as tired as the arm muscles that had to cope with heavy steering.

Dust was also a major problem, not just for the poor driver who was in the midst of it all, but also in keeping it out of the engine and radiator as the chaff screens and air-cleaners were not all that efficient. The tractors ran very hot, which caused a problem with the petrol engines that pre-ignited so much that it was nearly impossible to turn them off. Another disadvantage was the time it took to mount the

RIGHT: *The Ferguson FE-35 was based on the American TO-35 tractor but was fitted with Standard's engines and had a new style of hood. This is the diesel model and its 23C engine featured a Ricardo head.*

combine mechanism on the tractor, although the company claimed two men 'equipped with spanners and pliers' could fit it in twenty-five minutes.

A press announcement for the Ferguson combine was released in September 1956, stating that production was planned to provide delivery in time for the following year's harvest. However, the Ferguson design was a casualty of the rationalisation of the red and grey lines and was dropped in favour of the self-propelled Massey-Harris 735 that had been launched in 1956. The Ferguson combine project is believed to have cost the company over £1 million and yet came to nothing. Whether it would have met with any commercial success had it been put into production, we shall never know. Unfortunately, the 735 was a troublesome machine that never lived up to its expectations and was completely overshadowed by its larger brother, the far superior Massey-Harris 780.

At the same time as the combine programme, a lot of experimentation was carried out with sugar beet equipment. A rowcrop thinner for mechanically gapping beet plants, developed in conjunction with the National Institute of Agricultural Engineering at Silsoe, was launched in 1954. Ferguson single- and two-row beet lifters, based on the toolbar frame, were also offered from 1956.

The most elaborate piece of Ferguson beet equipment was a prototype two-stage harvesting system that was tested during the 1950s. It was designed by Horace Howell and was built in the engineering department at Fletchamstead. The first-stage unit consisted of a tractor fitted with a side-mounted topper and a rear-mounted top-saver that gathered the beet tops and elevated them into a cart so that they could be used for stock-feed. The second-stage lifter-loader was also a mounted unit and was carried on the tractor's rear linkage.

The Ferguson sugar beet harvester had two

BELOW: *The internal layout of the new hydraulic system fitted to the Ferguson FE-35.*

seasons on trial in the UK. Colin Steventon remembers spending one year testing the rowcrop thinner through the summer and the beet harvester through the winter. The harvester worked well in good conditions, but could not cope with wet or heavy land. The lifter cleaned the beet better after Dick Dowdeswell made some modifications to the shares and back rollers in the workshops of the local Ferguson dealers, Boston Tractors, while working with the harvester in South Lincolnshire.

It soon became evident that the Ferguson harvester would be unable to compete against the latest trailed machines coming on the market from manufacturers such as Peter Standen, Catchpole and GBW. As a last resort, the beet harvester was demonstrated in Yugoslavia, but it was not suited to the local conditions and performed badly. Consequently, Hermann Klemm cancelled the project, although the topper mechanism was incorporated into the rear-mounted Ferguson beet topper that was made in limited numbers from 1956.

Other items of Ferguson equipment that were developed but not put into production included a steerage hoe for maize, a single-row maize harvester and a bale thrower. A version of the American side-mounted forager harvester was also tested in the UK but was unable to handle the heavy and sometimes damp crops of grass grown in Britain.

A close scrutiny of some of the diesel tractors

ABOVE:
Nigel Liney takes the wheel for this publicity shot of a Ferguson FE-35. Introduced in 1956, the FE-35 was finished in a new grey and bronze livery.

used on test with the prototype combine and beet harvesters would have revealed a number of engine features that should have been out of place on a TE-20 model, including a rotary injection pump. The truth was that the engineers were developing a new engine in readiness for the new tractor that was to replace the TE-20. The field-test engineers had been testing a new cylinder head for the diesel engine for some time. The Ricardo head was designed by Sir Harry Ricardo and incorporated his Comet combustion chambers. The Comet indirect-injection combustion system was originally developed in 1931 for AEC to use in London buses.

BELOW: *The final assembly line for the FE-35 at Banner Lane. Standard spent over $4.5 million in re-tooling the Coventry plant ready for production of the new tractor to begin in September 1956.*

The head was eventually adopted for the new diesel engine that became known as the Standard 23C. It was based on the old 20C compression-ignition unit, but had a larger 84.14 mm bore giving it a capacity of 2,258 cc (137.8 cu in.). It had a CAV DPA-type rotary fuel injection pump with a mechanical governor. The compression ratio was raised to 20:1 and the engine developed over 37 bhp. The spark-ignition engines for the new tractor were also enlarged to 2,186 cc (133.4 cu in.) by boring them out to 87 mm. This gave the petrol engine 37 bhp, while the vaporising oil and lamp oil units were rated at 30 and 29 bhp respectively.

The replacement for the TE-20 was known as the Ferguson FE-35. It was based on Hermann Klemm's TO-35, and was mechanically identical apart from the Standard engines and the use of British components such as the Burman steering box and Lucas electrical equipment. It had the same advanced hydraulic system that incorporated implement transport, draft control, response control, position control, overload release and external hydraulic functions controlled by two levers. The lever on the outer quadrant controlled draft and set the working depth of the implement. The larger lever on the inner quadrant nearest the driver's seat was known as the operational lever. This lever provided position control in the upper sector of the quadrant, and controlled speed of response in the lower sector.

Where the FE-35 differed most from the TO-35 was in the sheet metalwork. The British tractor sported a more modern design of hood with a hinged top-panel for access to the battery, radiator cap and larger fuel tank, and a removable front grille for cleaning the radiator core. It also had a revised instrument panel housing an ammeter, oil pressure gauge

LEFT:
Lord Leigh in the seat for the launch of the Ferguson FE-35 that was held on 2 October 1956 at showrooms on Fletchampstead Highway.

or temperature gauge depending on which engine was fitted.

The FE-35 performed well on test, but its gear ratios, which were originally designed to match the North American Continental engine, did not really suit the Standard 23C diesel that was a little low on torque. However, the new hydraulics were more versatile, accurate and sensitive than the old TE-20 system, and a longer wheelbase and slightly heavier weight meant that the tractor was well balanced and could handle larger implements, such as a three-furrow plough, with ease.

Plans were made to put the FE-35 into

LEFT:
The vaporising oil version of the Ferguson FE-35. Powered by Standard's 87 mm engine, it developed 30 bhp. The tractor is seen working in Devon in the hands of collector Mike Thorne, with an FE-79 mid-mounted mower that was introduced in 1957.

RIGHT:
A petrol model Ferguson FE-35 tractor collecting milk churns from the dairy with a transport box.

production, and the Standard Motor Company announced in March 1956 that it was temporarily laying-off 2,500 workers for the summer while a new automated assembly line was installed at Banner Lane in readiness for the new tractor model. 'Laying-off' and 'automation' are not words that sit too well with factory workers, and the reorganisation caused a certain amount of industrial unrest. The dispute was settled when it became evident that there were not going to be mass redundancies and that the increased production at Banner Lane was going to bring greater job opportunities and higher wages.

Standard spent in excess of £4.5 million in equipping the Coventry plant for the increased mass-production of the FE-35 tractor. Just 150 of the latest modern and larger transfer machines, some weighing over 200 tons, replaced over 850 older machine tools, while about 1,000 other production and assembly machines had to be completely re-tooled. The company claimed that Banner Lane was now the largest and most modern tractor plant in the world and an annual production target was set at 100,000 tractors a year.

The last TE-20 was made in October 1956. A total of 517,651 had been built and

RIGHT:
A Ferguson FE-35 diesel with the spinner broadcaster that was also introduced in 1956. The de-luxe version of the tractor had a dual-clutch, 'live' power take-off, a tractormeter and a cushioned foam seat.

LEFT:
The Ferguson FE-35 industrial tractor was introduced in 1957. This vaporising oil model has the spring-cushioned bumper, dual-braking system, rear-view mirror, full-width fenders and road tyres.

despatched to all four-corners of the globe. It was the end of the era of the 'little grey Fergie' and the beginning of the last chapter in the Ferguson story as the FE-35 was launched on the market. In an attempt to move away from the all-grey Ferguson livery, several colour schemes were tried out on the new model before it was decided that the sheet metalwork and wheels would remain grey, while the engine and chassis would be finished in a distinctive metallic copper paint. Ferguson sales literature rather grandly called it a bronze finish, but the tractor is usually referred to as the 'grey and gold Fergie'.

The FE-35 went into production in September 1956 and was officially unveiled on 2 October at showrooms on Fletchampstead Highway belonging to the local Ferguson agent, S. H. Newsome. Standard or de-luxe model tractors were available with all four engine options offered, although the lamp oil version was still for export only. The de-luxe model had the dual-clutch giving 'live' power take-off and hydraulics, a tractormeter and a cushioned foam seat. Prices ranged from £463 for the standard petrol model to £590 for the de-luxe diesel tractor.

Between its introduction and June 1957, 40,000 FE-35s left Banner Lane. The average daily production was exceeding 360 tractors a day, and accounted for nearly 50 per cent of all the wheeled tractors produced in the UK. Export demand was exceeding supply, and 66 per cent of the total number of tractors built were shipped overseas to 142 different countries.

Industrial and vineyard versions of the FE-35 were introduced during 1957. The industrial model had a dual-braking system and could be equipped with full-width mudguards and industrial tyres. New Ferguson implements released that year included the FE-79 mid-mounted mower and the FE-93 range of mouldboard ploughs. These were some of the last machines to be marketed solely under the Ferguson name, as 1957 was the last year that the Ferguson and Massey-Harris product lines were kept separate.

The logistics of running two separate product lines was putting a financial strain on Massey-Harris-Ferguson, which lost over $4.7 million in 1957. James Duncan had been asked to resign in 1955, and Albert Thornbrough had been appointed as president in February 1956. Thornbrough was an able businessman who

ABOVE: *A diesel model FE-35 industrial tractor with the Ferguson fork-lift attachment that was introduced in 1956.*

recognised that steps had to be taken to rationalise and unite the two product lines if the company were to be turned around and become profitable.

In December 1957, Massey-Harris-Ferguson Ltd.'s shareholders sanctioned the re-naming of the company as Massey Ferguson Ltd. All the machines for 1958 were to be badged as Massey Ferguson equipment. Tractors were to be painted red and grey, all mounted and Ferguson System equipment would remain grey, while trailed implements and combines would be red. There would be one line of Massey Ferguson distributors and dealers, and the company's products would be identified by a new triple-triangle symbol.

The Ferguson FE-35 was re-launched at the December Smithfield Show as the Massey Ferguson 35 tractor in the red and grey livery. It was joined by a new larger tractor, the Massey Ferguson 65. The 65 was based on the Ferguson 40 concept. It differed from the North American 65 in that it had Ferguson-influenced styling and was only available as a diesel. It was powered by a four-cylinder Perkins 4.192 engine, a 3,146cc (192 cu in.) power unit developing 50 bhp.

This use of a Perkins engine was a significant move for Massey Ferguson. The demand for petrol and vaporising engines was falling, and most 35 tractors supplied were diesels. The problem was that the Standard 23C diesel was not the best compression-ignition engine on the market and was not suited to the FE-35. It was a poor starter and the 35 tractor was in danger of becoming outclassed by competitors' machines.

Neither were the Standard engines that cheap, and the relationship between Massey Ferguson and the Standard Motor Company became strained after the two firms clashed over the cost of producing the tractors. Thornbrough felt that Standard was holding Massey Ferguson to ransom and was determined that his company

LEFT:
A staged publicity shot illustrating all the advantages that Massey-Harris-Ferguson offered to its customers. The photograph was taken at Stoneleigh using training instructors and staff from engineering.

Key:

(1) dealer and personnel training;
(2) design and engineering;
(3) the dealer network;
(4) service in the field;
(5) spare part replacement scheme;
(6) finance scheme;
(7) re-sale value;
(8) the FE-35 tractor;
(9) the equipment range.

should secure its own engine and tractor manufacturing facilities to reduce its dependence on outside suppliers.

Things were looking better for the company by 1958. The old Ferguson tractor plant in Detroit that had been re-designed and extended was re-opened on 7 May. The company also purchased the Borg-Warner transmission plant that had supplied axles and gearboxes for the North American tractors. Sales were up, and Massey Ferguson's annual turnover for 1958 exceeded $440 million, leaving it a pre-tax profit of nearly $22 million. With money coming into the bank, Thornbrough had a shopping list and wanted to acquire both the Perkins company and Standard's tractor plants, including Banner Lane.

The Perkins deal was concluded in January 1959 after the board of the Peterborough company recommended its shareholders to

LEFT:
This photograph, taken on the old army parade ground near the river at Stoneleigh, shows training school instructors lined up with both Ferguson and Massey-Harris equipment. FE-35 tractors are in the front line, while on the back row are (left to right) a 780 combine on rice tracks, a bagger-model 780 combine, a 735 combine, a 55K tractor and three 745 tractors.

RIGHT:
In December 1957, the Massey-Harris-Ferguson company was renamed Massey Ferguson, and the Ferguson FE-35 was re-launched as the Massey Ferguson 35 in a new red and grey livery. This diesel model is seen on trial with the 718 automatic potato planter near Stoneleigh in early 1958.

accept Massey Ferguson's offer, and F. Perkins Ltd. was acquired for almost £4.5 million. The negotiations with Standard were more protracted. It appeared that Massey Ferguson had being buying up Standard shares since early 1957 in an attempt to win control of its tractor production and by the time of the Perkins buyout it owned nearly a quarter of the Standard Motor Company's shares.

Standard was keen to sell its tractor interests to finance other projects, including enlarging the Canley plant and building a new assembly area for the Triumph Herald car, but was not prepared to be pressurised by the giant Canadian corporation. It still had seven years left in its contract to build Ferguson tractors, and believing itself to be in a good bargaining position, asked £13 million for its tractor facilities, which was far more than Massey Ferguson was prepared to pay. This resulted in a game of brinkmanship between the two companies.

In February 1959, Massey Ferguson announced that if an agreement could not be reached with Standard then it intended to buy or build another tractor plant elsewhere. It also stated that it would not be renewing the contract with the Coventry firm that would run out in 1966. In March, the Canadian organisation said that it was ready to switch over almost totally from Standard to Perkins diesel engines. The Standard Motor Company countered this by indicating that it might begin producing its own range of tractors.

Standard had in fact been experimenting with tractor designs for some time and in 1958 the company built two prototype machines, designed by John Chambers and Trevor Knox in conjunction with Alick Dick. Chambers and Knox had left Massey-Harris-Ferguson to join the Standard tractor project in

RIGHT:
The British Massey Ferguson 65 was launched in December 1957. Based on the Ferguson 40, it was powered by a Perkins 4.192 diesel engine developing 50 bhp.

LEFT:
One of two surviving Standard prototype tractors. Built in 1958, it is fitted with the company's 23C diesel engine. This example now forms part of the Coldridge Collection in Devon.

protest against Hermann Klemm's dominance of MHF engineering.

Details of the Standard project are sketchy, but it appears that it was badged as the SMC tractor and was very similar in appearance to the Ferguson. It was fitted with the 23C diesel engine, but had a different cylinder head with separate heater plugs in each combustion chamber. The transmission consisted of a four-speed gearbox mated to a two-speed epicyclic reduction unit, both controlled by a single gear lever operating two sets of selectors. This gave eight forward and two reverse speeds. A lever to the right of the seat had two separate gates for draft and position control hydraulics, and the tractor had inboard disc brakes and 'live' power take-off. Several automotive components were used, including a fuel tank from the Standard Atlas van.

It seems that a second pair of SMC prototypes followed, both slightly different to the first two tractors, and fitted with a forward and reverse shuttle transmission and more modern sheet metal. Two examples of the Standard tractors have survived into preservation, one of each version.

The impetus to produce a Standard tractor waned after Banner Lane was finally sold to Massey Ferguson. Work continued for a time, but the SMC project was quietly dropped after Standard-Triumph was taken over by Leyland in 1961, while the improved Standard 23C engine was supplied to Allis-Chalmers for its new British diesel tractor, the ED-40, that was produced at Essendine from 1960.

Standard's deal with Massey Ferguson had

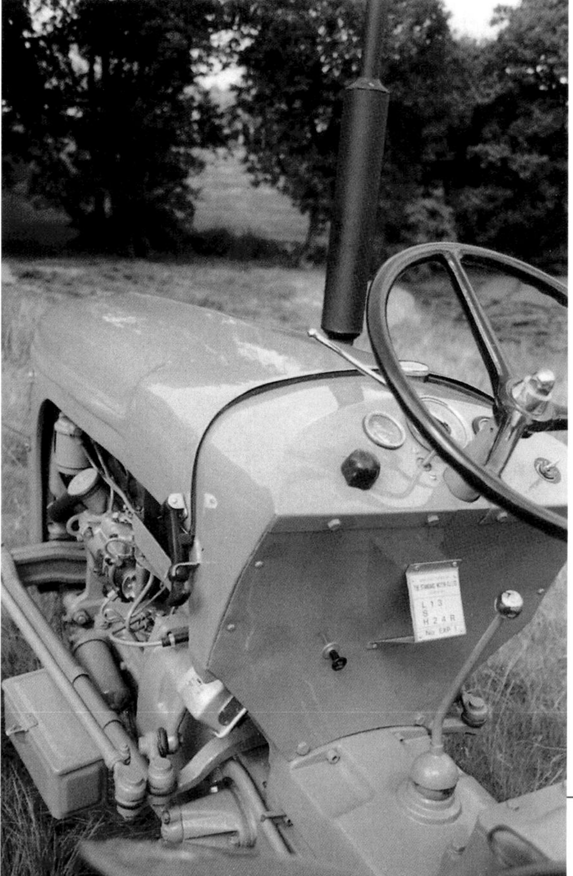

LEFT:
The first Standard prototype tractors had a four-speed gearbox mated to a two-speed epicyclic unit with a single gear lever operating two sets of selectors giving eight forward speeds.

RIGHT:
A second and slightly different Standard prototype tractor. Owned by Robert Crawford of Boston in Lincolnshire, it has modern sheet metal and would have been marketed as the SMC tractor had it been put into production.

been finally concluded in July 1959, and the Canadian organisation purchased all the Standard Motor Company's tractor assets for £11.8 million plus its Standard shares. The acquisition included Banner Lane and the French facilities at Beauvais and St. Denis. The Canley engine plant, which was not needed with the recent takeover of Perkins, was not included in the sale. However, Standard agreed to manufacture the spark-ignition engines for the 35 tractor for the foreseeable future, as well as a small quantity of diesel engines to tide over the transition to Perkins power units.

The 35 tractor with the three-cylinder Perkins 3A-152 diesel engine was launched at the end of 1959 at the Smithfield Show. It was a much better tractor than the previous diesel model as the torque characteristics of the Perkins engine were similar to those of the Continental Z-134 that had powered the North American 35 and the best use could now be made of the gear ratios. The 35 and 65 tractors that were rolling out of Banner Lane at the rate of 400 per day were the first of a long line of Massey Ferguson tractors to be built at the Coventry plant. It built its three millionth tractor in 1996, and the factory still accounts for a major part of Massey Ferguson's worldwide tractor production.

Today, Massey Ferguson is part of the giant AGCO Corporation, based in Duluth in Georgia. Beauvais remains an important facility, and builds the larger MF tractor models. In 1999, it took over the manufacture of the high-horsepower tractors sold under the Agco-Allis and White brand names after AGCO transferred production from its Coldwater plant in Ohio.

BELOW:
The gear levers on the second prototype Standard tractor were arranged differently to accommodate a shuttle transmission. The back-ends of the Standard tractors were highly engineered and would have coped with a far more powerful engine than was actually fitted.

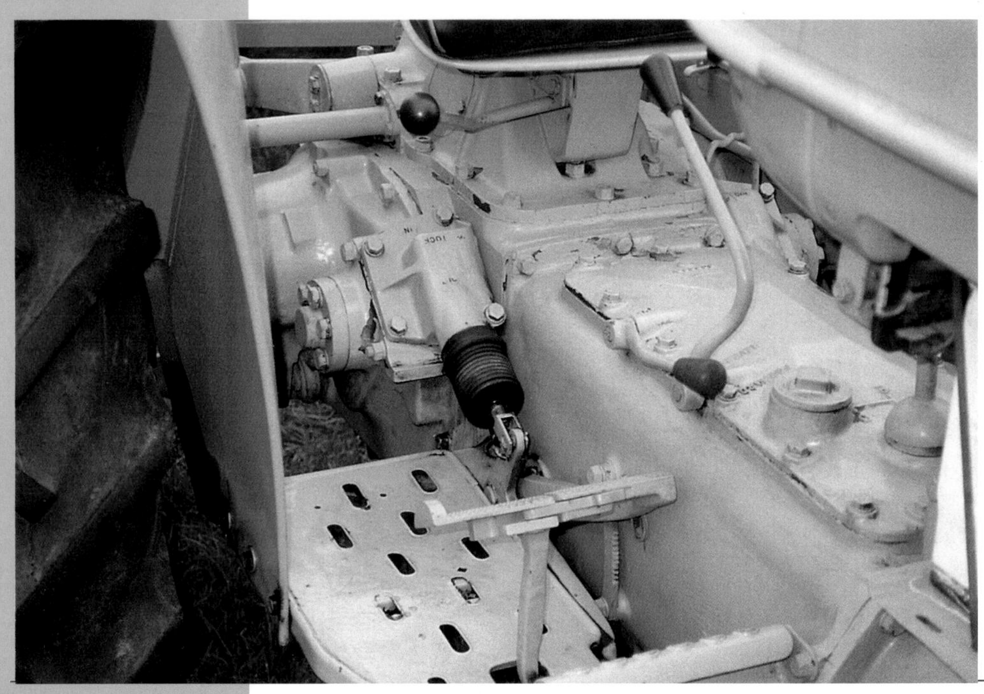

LEFT:
Tony Rolt demonstrates the R2 prototype Ferguson car at Abbottswood in 1952. The vehicle had four-wheel drive and its flat-four engine was mounted at the rear.

Epilogue
The Ferguson Legacy

Even though he was nearly seventy years old at the time of his resignation from MHF in July 1954, Harry Ferguson was not prepared to relax into retirement. His mind was too active for that and he had other irons in the fire, including a number of automotive developments carried out in the name of a separate company, Harry Ferguson Research Ltd.

Ferguson's passion for motoring had been rekindled in 1948 when he agreed to back the development of a four-wheel drive car known as the Crab. This Riley-powered machine also steered on all four wheels, and was the brainchild of two racing drivers, Freddie Dixon, an old associate of Ferguson's, and Tony Rolt.

As the project progressed, Ferguson brought in an ex-Aston Martin engineer, Claude Hill, to join the team that was based at Redhill in Surrey. In April 1950, he formed Harry Ferguson Research Ltd. with the intention of building a 'safety' family car with four-wheel-drive, moving the company into a corner of Banner Lane two years later. Dixon, however, felt Ferguson had hijacked his project; he became disenchanted with the whole affair and left the company soon afterwards.

Experiments over the next few years saw the original Crab design discarded and replaced by the more refined R2 and R3 prototypes. The company introduced several advanced safety features, and pioneered a number of unique developments. These included the Ferguson-Teramala transmission, a hydraulic torque converter that was mated to a normal gearbox and acted as a semi-automatic transmission, and the Ferguson-Maxaret anti-lock braking system that was developed in conjunction with Dunlop. Claude Hill was working on a horizontally-opposed four-cylinder engine and a controlled differential.

Harry Ferguson took a more active interest in the designs after his break from Massey-Harris-Ferguson, and saw all the features brought together as a road-going car in the R4 prototype completed in 1954. The company could no longer stay at Banner Lane, and after finding temporary accommodation at Chipping Warden aerodrome in Oxfordshire, it moved to a permanent location at Toll Bar End in Coventry in July 1956.

The R5 estate car appeared in 1959, but commercial success still eluded the company as no major manufacturer seemed prepared to take on all or part of the designs. Ironically, many of the advanced safety features of Ferguson's cars, including four-wheel drive, traction control, anti-lock braking, seat belts and collapsible steering columns, are taken for granted today, but at the time they were probably regarded as expensive gadgetry. With less complicated

RIGHT:
A prototype Ferguson cross-country vehicle based on the four-wheel drive R2 car. It is seen on trial at Abbottswood, climbing the banks of the River Dikler.

specifications, the designs might have found their way into volume production, but Harry Ferguson was never one to compromise his idealism for the sake of commercialism; his objective was safer motoring and nothing would sway him from that vision in the same way that he had doggedly hung on to his dream of improving mechanised farming.

Harry Ferguson's first love was still agricultural engineering, and he could not completely distance himself from tractor developments. When he left Massey-Harris-Ferguson, he managed to extract an agreement from the board that would allow him to return to tractor production under his own name after five years had elapsed, and he had the notion to build a new lightweight machine.

In 1956, he began planning the new tractor and even settled on a price for it of £400. He established the tractor research division as part of Harry Ferguson Research, and in 1960 he persuaded Alex Senkowski to join the company to head the design team. The brief was to build a machine weighing less than a ton and incorporating the Ferguson-Teramala transmission.

The main thrust of the company's activities for 1960, however, was the development of a racing car to publicise the Ferguson four-wheel drive system. Codenamed Project 99, the four-wheel drive racer was developed at Harry

BELOW:
Project 99 - the Ferguson four-wheel drive racing car. Unveiled in 1960, it was powered by a 2.5-litre Coventry Climax engine.

LEFT:
The prototype lightweight tractor that was developed by Harry Ferguson Research Ltd. on test in 1962.

Ferguson's instigation. Sadly, he was to see neither the tractor nor the P99 car project come to fruition as he suddenly passed away while taking a bath on 25 October 1960.

Harry Ferguson left behind him a legacy of achievement that few other mortals can match. He revolutionised agriculture, pioneered in aviation and saw through some remarkable automotive developments. Although he did not live long enough to see Harry Ferguson Research achieve the success it deserved, the company was on the brink of public and commercial recognition.

The Ferguson P99 racing car made its debut at Silverstone on 8 July 1961. Interestingly, it was the first Ferguson machine to be fitted with a Coventry Climax engine since the Ferguson-Brown tractor of 1936. The cubic capacity was similar at 2.5 litres, but there all similarity ended. This was a highly tuned, overhead-valve, lightweight racing engine and it pushed the P99 to victory in the Oulton Park Gold Cup race the following September with Stirling Moss at the wheel.

The interest that the racing car aroused was just the publicity that Harry Ferguson Research needed. The motoring press clamoured to test-drive the Ferguson cars, and the company received inquiries from several manufacturers for its developments. A deal was struck with Jensen Motors of West Bromwich who wanted to incorporate the four-wheel drive and anti-lock braking systems in a version of its Interceptor high-performance car. The car was launched in 1967 as the Jensen FF, standing for 'Ferguson Formula'. Harry Ferguson Research continues today as FF Developments and is part of the Ricardo Group. It is run by Tony Rolt's son, Stuart, and has been responsible for many important automotive innovations over the years.

BELOW:
Designed by Alex Senkowski, the prototype lightweight tractor was only a 15 bhp machine and was powered by a four-cylinder diesel engine based on the BMC A-series power unit.

RIGHT: *The BMC Mini tractor that was launched in December 1965. It was developed by Harry Ferguson Research Ltd. from a project that began in 1960. Note the typical Ferguson steering and front axle arrangement.*

The lightweight tractor project also became viable after the Nuffield Organisation approached Tractor Research in 1960, wanting a new small model to add to its range. An agreement was reached by the end of the year, and work started on re-designing the little tractor to meet Nuffield's requirements. The Ferguson-Teramala transmission was dropped in favour of a three-speed constant-mesh gearbox with an integral three-range splitter-box giving a total of nine forward and three reverse gears.

The tractor took three years to develop, the first prototype bing ready for field trials by 1962. The lightweight machine was very much like a scaled-down TE-20, and the steering and front axle arrangement were pure Ferguson. Nuffield was part of the British Motor Corporation and the engine was sourced from within the group. Developed with assistance from Harry Ricardo's company based in Shoreham, it was a diesel version of the famous BMC A-series petrol engine that had already powered over 2 million Austin and Morris cars, trucks and vans. The four-cylinder tractor engine was a diminutive unit with a cubic capacity of only 948 cc (57.9 cu in.). It developed 15 bhp and its cylinder head incorporated Ricardo Comet V combustion chambers.

Nuffield launched the tractor in December 1965 as the BMC Mini. It weighed just over 2,000 lb and cost £585. Specification included power take-off, disc brakes and a hydraulic lift with mechanical draft control and weight transference.

Was this little orange machine the last Ferguson tractor? It may have incorporated many of Harry Ferguson's principles and matched the criteria he laid down for his last lightweight model, but whether it was the embodiment of his final vision, we can only guess. His supreme achievement must surely be the countless thousands of Ferguson tractors that are still working or are in preservation worldwide, and his legacy the hydraulic system and linkage that made them famous - a system that remains at the heart of most of today's tractors.

Appendix 1
Ferguson Conversions

The agricultural tractor, with its simple design and rugged construction, was a very adaptable power unit and was used as the basis of a considerable number of specialist machines for industrial, construction or agricultural use. The manufacturers of these machines depended on a supply of skid-units, so-called because the tractor was normally supplied less-wheels on a wooden skid, from the better-known tractor concerns. The advantage for the specialist manufacturer was that they got an inexpensive and proven power unit, and could benefit from the tractor company's established spares and dealer network.

ABOVE:
This Ford 2NAN model was a Perkins experimental tractor and was fitted with a P4 diesel engine in 1945. It is seen working on Frank Perkins' own farm at Alwalton near Peterborough. Perkins also fabricated the steel cab and engine side-panels.

During the 1950s and 1960s, Ford dominated this market in the UK with a dependable product that had worldwide recognition and back-up. Ferguson tractors were as universally accepted, but their lighter construction and lower power meant that they were not as popular with most of the construction or industrial machinery manufacturers and Ferguson conversions were not so common.

Because the earlier Ferguson tractors were limited by power and the lack of a diesel engine option, they were ideal candidates for Perkins diesel conversions. Frank Perkins began building compression-ignition engines in Peterborough in 1932, introducing his famous P-series power units in 1937. In May 1945, he experimented with fitting a P4 engine to an imported Ford Ferguson 2NAN model. The four-cylinder Perkins P4 developed 30 hp, and the tractor was used successfully for several years on Frank's own farm, using only one gallon of diesel fuel per acre when ploughing with a three-furrow plough. It was really only an evaluation exercise and no plans were made to market the conversion; it was judged to be too expensive, the engine was really too large for the tractor and some concern was expressed over the amount of torque it would put through the rear transmission.

The answer for the Ferguson was the later and smaller three-cylinder P3 (TA) agricultural diesel engine that was brought out by Perkins in the

LEFT:
The Perkins conversion for the Ferguson TE-20 was launched in 1950 and used the P3 three-cylinder diesel engine. It was an economical and reliable power unit that developed 32 bhp.

RIGHT: *A Ferguson tractor with the Perkins P3 conversion harrowing peas on C. D. Tebbs' farm near Peterborough in 1951.*

early 1950s. This 32 bhp unit slotted neatly into the Ferguson in place of the old petrol or vaporising oil engine, offering increased power, greater economy and better torque at lower speeds.

There was still a demand for the Perkins conversion even after Ferguson brought out its own diesel tractor, the TE-F20 with the Standard Motor Company's 20C engine. This was because the Standard diesel could not be fitted to the earlier petrol and vaporising oil tractors without extensive modifications to the transmission. The Perkins diesel, although more expensive, was probably also a better engine; it had 4 bhp more power than the Standard and was much easier to start.

The first Ferguson to be fitted with a Perkins P3 was a TE-20 that was used for field trials in September 1950, ploughing up Ferry Meadows, now a beauty spot and an important recreation area in Peterborough. The P3 conversion was launched in November 1950 and was available for the TE-20, Ford Ferguson and Ford 8N tractors. Perkins sold it as a conversion pack with all the necessary parts to fit the P3 into the tractor, including an adapter plate to mate the engine to the bell-housing. Because of the height of the P3, the hood had to be raised with a new metal insert fitted over the dashboard, giving the Ferguson tractors with Perkins engines a very distinctive profile.

A considerable number of Perkins conversions were sold, particularly for the petrol Fergusons that were becoming expensive to run. The later

RIGHT: *The Perkins conversion pack for the TE-20 included the engine and an adapter plate for the bell housing. Note how the height of the hood had to be raised to fit over the engine. The tractor is seen using a Hydrovane 25 compressor to power the hedge-cutter attachment.*

LEFT:
This Perkins development tractor was fitted with an A4.107 four-cylinder diesel engine in July 1961. It is believed that the tractor was being used as a test bed to evaluate the engine for the Massey Ferguson 130.

P3/144 engine was introduced as an improved conversion for the Ferguson from April 1957, and Perkins continued to experiment with other engine options. In July 1961, Perkins fitted a TE-20 with the four-cylinder A4.107 engine that had been used in the French-built Massey-Harris 25 from 1959. It seems that the object of the exercise was to use the tractor as a test-bed to evaluate the engine for the Massey Ferguson 130, the model that replaced the 25 in 1964, rather than as a genuine Ferguson conversion. What is believed to be a second prototype of the Ferguson with the A4.107 engine has recently come to light and is now in preservation.

Other Ferguson conversions usually took the form of simple adaptations for specialist applications. Several of the larger tractor manufacturers, including Ferguson and Ford, preferred to concentrate on a standard wheeled agricultural machine for high-volume production. Industrial versions were offered, but they were basic tractors requiring little extra modification. For more specialist applications, customers had to turn to one of the smaller outside companies who were offering conversion kits or extra equipment. Many of the conversions were often

LEFT:
A Ferguson TE-20 with the Perkins A4.107 engine. This prototype machine has only recently come to light.

RIGHT: *The Bryden Tracpak fitted to a Ferguson TE-D20 vaporising oil tractor. Introduced in 1951, the Tracpak full-track conversion was designed by George Bryden of Leeds.*

supplied by dealerships to suit the particular needs of their local area.

As already mentioned, the Ferguson was ideally suited to hillside work, but if extra traction were needed, there were several track conversions on the market. One of the most interesting, the Bryden Tracpak, turned the Ferguson into a full crawler with proper steel tracks. The conversion was devised by George Bryden Engineering of Seacroft, Leeds. Bryden had been managing director of Marshall's of Gainsborough and was involved with the development of the Fowler Challenger crawlers. He designed the Tracpak in conjunction with his eldest son.

The Tracpak consisted of two track-frame assemblies, each with three pairs of rollers and

RIGHT: *A Bryden Tracpak conversion fitted with a Perkins P3 diesel engine. This tractor is seen on F. A. Standen's stand at the Peterborough Show in July 1952.*

LEFT:
The Perkins-powered Bryden Tracpak at Ferry Farm, Sudbourne, near Ipswich. The tractor, equipped with a dozer blade, is clearing flood damage in February 1953.

two idlers mounted on spring-loaded suspension units. The track units were attached by a crossbeam to the transmission and brackets to the Ferguson's front axle bar. The 10 in. track had inverted teeth driven by the sprockets that were fitted in place of the tractor's rear wheels. The crawler was steered by the steering wheel with the steering arms reversed and coupled to the brake rods through swivel links. Two men could fit the Tracpak in two hours, and the company claimed it exerted a ground pressure of only 5 psi.

The Bryden conversion was introduced at a field demonstration held at Well Green Farm near Garforth in Yorkshire on 4 September 1951. A Tracpak fitted to a petrol TE-20 was shown ploughing, cultivating and driving a binder on a 1 in 6 incline. The machine created

LEFT:
A Ford Ferguson 2N model fitted with Roadless DG13 half-tracks in 1944.

ABOVE: *A Ferguson TE-20 with the Perkins P3 conversion and the Selene four-wheel drive system that was developed in Italy in 1953.*

a lot of interest, but very few were sold.

In July 1952, the Ferguson dealers, F. A. Standen of Ely, exhibited a Tracpak conversion fitted with a Perkins P3 diesel engine at the Peterborough Show. This tractor was sold to Ferry Farm at Sudbourne, near Ipswich, and was equipped with a dozer blade. On 3 February of the following year, hurricane-force winds combined with high tides to swamp England's east-coast flood defences. East Anglia was badly affected and nearly half of Ferry Farm's 1,000 acres were under water. The Tracpak Ferguson crawler was pressed into service clearing up and was one of the few machines light enough to work on the water-logged land.

A similar Ferguson crawler conversion was made by a Somerset farmer, Mr L. C. Burdge of Yatton, in June 1951, using tracks, front axle and idler wheels from an ex-War Department amphibious vehicle. Another full-track conversion for the Ferguson was marketed by H. Cameron Gardner Ltd. of Reading in Berkshire. This consisted of a flexible track that wrapped around the tractor's tyres. Again, the brakes were used for steering, but this time they were cable-operated from the steering wheel.

Half-track conversions included Ferguson's own based on the Canadian Bombardier design, and the heavier steel-girder Roadless track units. Roadless Traction of Hounslow introduced its Driven Girder, or DG, half-track in 1944. Designed to lock in a predetermined curve equal to a wheel of 20 ft diameter, it was offered as a conversion for several makes of tractor into the 1950s. A DG13 lighter version of the half-track was available to suit the Ford-Ferguson, Ford 8N and Ferguson TE-20, as well as the Allis-Chalmers Model B tractors.

A four-wheel drive system for the Ferguson TE-20 was developed in 1953 by the Italian Selene company based in Nichelino near Turin. The conversion, originally designed for hillside ploughing in the Po Valley area of Italy, consisted of a sandwich transfer box driving an American Jeep front axle. Selene offered the tractors as a complete unit fitted with the Perkins P3 engine, and a few of these were imported by Reekie Engineering of Arbroath in Scotland.

RIGHT: *The Reekie narrow or 'Berry' Tractor based on a Ferguson TE-D20. About 500 were made by Reekie Engineering between 1948 and 1952.*

Selene's chief engineer, Signor Torchio, went on to devise other systems, and the company introduced a four-wheel drive Ferguson with the transfer box driven off the rear ground-speed pto. This system was marketed in the UK by the Robert Eden company of London, a prominent exporter who had previously supplied second-hand tractors to Selene. Robert Eden also launched a Selene four-wheel drive version of the Ferguson FE-35 at the 1956 Bath Royal Show.

Before the vineyard TE-20 was introduced, the soft-fruit growers in the north-east of Scotland were catered for by Reekie Engineering. The founder of the company, John Reekie, had served with the Royal Mechanical & Electrical Engineers during the Second World War. In 1947, he set up a Ferguson dealership in partnership with his brother, Gavin, trading as Farm Mechanisation Ltd. He devised the Reekie conversion in 1948 after being approached by raspberry growers, including Chivers Jams, who found the narrow Ferguson too wide and needed a tractor with an overall width of no more than 40 in.

Reekie modified the tractors on his own premises at Lochlands Works in Arbroath. The conversion involved twenty-eight separate alterations, including substantial modifications to the front-axle layout. The rear-axle half-shafts were also cut and re-welded. The track was adjustable in 4 in. steps and could be altered from 32 in. for fruit or hop work to 56 in. for normal operations.

The narrow-track Fergusons were known locally as 'Berry Tractors'. The conversion was launched in June 1949 and cost £250. Reekie also offered special implements to match the tractor, including raspberry ploughs, discs, fruit sprayers and a narrow trailer. Around 500 were sold, but demand for the conversion dwindled after Ferguson launched its own vineyard model in 1952.

Lenfield Engineering, which had been involved with Ferguson's own narrow tractor since 1948, built a few lesser-known conversions for working in hop fields. This Kent concern also made a unique modification to a Ferguson for spraying blackcurrant bushes. The tractor was mounted on stilts with drop-boxes enclosing a

LEFT: *A 1950 Reekie narrow conversion of the Ferguson tractor. The modification was launched at the request of raspberry growers who wanted a tractor that was no more than 40 in. wide.*

BELOW: *A 1947 Ferguson TE-20, fitted with a Continental petrol engine, that had been turned into a high-clearance sprayer by Lenfield Engineering of Kent. The drop-boxes to the rear wheels enclosed a chain drive from the rear axle.*

RIGHT: *The Lenfield high-clearance sprayer conversion of the Ferguson tractor at work. It was designed for spraying blackcurrant bushes.*

chain drive from the rear axle to the rear wheels.

Several other dealers converted Ferguson tractors into high-clearance machines for spraying operations. Most of these were simple adaptations with extended stub-axles and modified fenders to accommodate larger-diameter rowcrop rear wheels. The sprayer manufacturers, J. W. Chafer of Doncaster, used a fleet of Fergusons converted in this way for their contract crop dusting and chemical spraying fleet.

In France, the Société de Motoculture du Gard from Nimes marketed a high-clearance version of the Ferguson for working in vineyard plantations. The Ferguson power unit was mounted on a separate rigid chassis known as the 'tractor bridge'. A chrome-nickel chain-transmission powered the rear wheels. Introduced in June 1954, the unit was capable of

RIGHT: *J.W. Chafer of Doncaster used a fleet of Ferguson tractors for contract crop spraying operations, including this 1948 TE-A20 petrol model with rowcrop wheels.*

LEFT:
This TE-F20 diesel tractor was one of several modified in J. W. Chafer's workshops at Doncaster for crop dusting. It featured extended stub-axles, modified fenders and larger-diameter rowcrop rear wheels.

straddling two rows of vines and had up to 4 ft 4 in. ground clearance.

Industrial conversions of the Ferguson were scarce; several were used as loading shovels and some were converted into turf-rollers for playing fields. In 1953, E. V. Twose Ltd. of Tiverton in Devon introduced the Tractamount roller, a simple road-roller conversion powered by a TE-20.The Ferguson was driven up onto the roller chassis, which it drove by a heavy-duty chain through special sprockets bolted to the tractor's rear wheel-centres.

Other strange conversions included the Borrow mobile drier, a rotary drum-type drying plant powered by a TE-20 and built by E. W. Borrow of Cowplain, Portsmouth in 1952. This is not a full list of Ferguson conversions, but there were few others; the tractor's real strength lay in what Harry Ferguson designed it to be – a simple but innovative agricultural machine.

LEFT:
Launched in 1953, the Twose Tractamount was a unique conversion to turn the Ferguson tractor into a road roller.

Appendix 2

Ferguson Tractor Serial Numbers & Designations

Ferguson Type A

No listings of serial numbers are available for the Type A, which was built from approximately April 1936 until June 1939.

Notes

1. Coventry Climax engine fitted up to serial No. 256. David Brown engine phased in between serial Nos. 256 & 500.
2. Last recorded Type A tractor serial No. 1354.

Ford 9N & 2N

Year	Serial No.	Year	Serial No.
1939	1	1944	126538
1940	10234	1945	169982
1941	45976	1946	198731
1942	88888	1947	258540
1943	105375		

Notes

1. First 9N built 5 June 1939.
2. First 2N, No. 99003, built April 1942
3. All 9N and 2N tractors produced in same serial number series, prefixed '9N'.

Ferguson TE-20

Year	Serial No.	Year	Serial No.
1946	1	1952	241336
1947	316	1953	310780
1948	20895	1954	367999
1949	77773	1955	428093
1950	116462	1956	488579
1951	197837		

Notes

1. All models in same numbering sequence.
2. First TE-20 built 6 July 1946
3. First TE-A20 serial No. 8710
4. First TE-D20 serial No. 94952
5. First TE-F20 serial No. 200001
6. Last TE-20 model, serial No. 517651, built October 1956

Ferguson TO-Series

Year	Serial No.	Year	Serial No.
1948	1	1952	72680
1949	1808	1953	108645
1950	14660	1954	125959
1951	39163		

Notes

1. TO-20 and TO-30 Series in same numbering sequence.
2. First TO-20 built 11 October 1948.
3. First TO-30, serial No. 60001, built August 1951.

Ferguson TO-35

Year	Serial No.	Year	Serial No.
1954	140001	1956	167157
1955	140006		
1957	171741		

Notes

1. First TO-35 built December 1954
2. Ferguson TO-35 became Massey Ferguson 35 from December 1957

Ferguson FE-35

Year	Serial No.	Year	Serial No.
1956	1001	1957	9226

Notes

1. First Ferguson FE-35 built September 1956
2. Last Ferguson FE-35 built December 1957. Tractor became Massey Ferguson 35 from that date.
3. Prefix: SKF = Standard VO, SKM = De Luxe VO, SDF = Standard Diesel, SDM = De Luxe Diesel.

General Note

Serial numbers are for first tractor built on the first working day of each year with the exception of model introductions. Lists have been compiled from Ferguson service bulletins, but no guarantee is given for their accuracy and they should be used as a guide only.

Ferguson TE-20 Model Designations

Model	Type	Fuel	Remarks
TE-20	Normal-width	Petrol	Continental engine
TE-A20	Normal-width	Petrol	
TE-B20	Normal-width	Petrol	Continental engine
TE-C20	Narrow	Petrol	
TE-D20	Normal-width	Vaporising oil	
TE-E20	Narrow	Vaporising oil	
TE-F20	Normal-width	Diesel	
TE-G20	Narrow	Diesel	Only made in France
TE-H20	Normal-width	Lamp oil	
TE-J20	Narrow	Lamp oil	
TE-K20	Vineyard	Petrol	
TE-L20	Vineyard	Vaporising oil	
TE-M20	Vineyard	Lamp oil	
TE-N20	Vineyard	Diesel	Only made in France
TE-P20	Industrial	Petrol	
TE-R20	Industrial	Vaporising oil	
TE-S20	Industrial	Lamp oil	
TE-T20	Industrial	Diesel	
TE-PT20	Semi-industrial	Petrol	
TE-TT20	Semi-industrial	Diesel	
TE-PZD20	Industrial	Petrol	Agricultural fenders
TE-PZE20	Industrial	Petrol	No fenders
TE-TZD20	Industrial	Diesel	Agricultural fenders
TE-TZE20	Industrial	Diesel	No fenders

Bibliography

JOURNALS & PERIODICALS

Aeroplane Monthly
Agricultural Machinery Journal
Country Life
Coventry Evening Telegraph
David Brown Tractor Club Journal
Farm Mechanisation
Farmers Weekly
Ferguson Club Journal
Ferguson Heritage
Implement & Farm Machinery Review
Overseas Engineer
Machinery
Mourne Observer
New Ulster
News Review
Power Farmer
Standard Car Review

BOOKS

Cherouvrier, Jean, and Noulin, Jean, **Tracteurs Ferguson** (ETAI, 1999)
Donneley, Desmond, **David Brown's - The Story of a Family Business** (Collins, 1960)
Farnworth, John, **Ferguson Implements & Accessories** (Farming Press, 1996)
Farnworth, John, **The Massey Legacy -Volume One** (Farming Press 1997)
Fraser, Colin, **Harry Ferguson - Inventor & Pioneer** (John Murray, 1972, Old Pond, 1998)
Fredriksen, Erik, **The Legendary LTX Tractor** (Erik Fredriksen, 2000)
Gibbard, Stuart, **The Ford Tractor Story - Part One** (Old Pond Publishing & Japonica Press 1998)
Leffingwell, Randy, **Ford Farm Tractors** (MBI Publishing Company, 1998)
Martin, Bill, **Harry Ferguson** (Ulster Folk & Transport Museum)
Rae, John B., **Harry Ferguson and Henry Ford** (Ulster Historical Foundation, 1980)
Reid, Hugh, **From Apprentice to Chairman** (Hugh Reid, 1993)
Sorensen, Charles E., and Williamson, Samuel T., **Forty Years with Ford** (Jonathan Cape, 1957)
Williams, Michael, **Massey Ferguson Tractors** (Blandford Press 1987)
Winter, Michael, **Harry Ferguson & I** (Michael Winter, 1995)
Wymer, Norman, **Harry Ferguson** (Phoenix House Ltd., 1961)

TECHNICAL PAPERS

Foxwell, W. J., **Draft Control Mechanisms** (Ford Motor Company, 1960)
Foxwell, W. J., **Hitches and Hydraulic Systems** (Garland Law Publishing, 1983)
Hockey, W. S., **Tractor Mounted Implements and Adaptations** (Institution of Mechanical Engineers, 1961)

Index

Roman numbers refer to text, bold to illustrations

A

AGCO Corporation, 146
Allis-Chalmers, 26, 145
    All-Crop combine, 52
Anderson, John, 30
Annat, Bob, **29**
Aspin, F.M. 90
Australian National Antarctic Research Expedition, 130

B

BMC Mini tractor, 150, 150
Ball, Frankie, 127
Balmer Lawrie Ltd, 106
Belkowski, George, 80, **126**, 135
Bibby, Jack, 78, 127, 128
Black, Sir John, 64-6, **65**, 68, 74, 122
Bofors anti-aircraft guns, 55
Boilliant-Linet, Jacques, 104
Borg-Warner transmission plant, 143
Borrow Mobile drier, 159
Botwood, Allan, 78, 121
Boyd-Brent, Peter, 106, 134
British Overseas Engineering & Construction Corporation, 99
Brock, Harold, 49
Brown, David, 34, 39, 46, 48
Brown, Frank, 34
Bunting, Ernest, 110, 114

C

Chafer, J.W, 158, 159
Chambers, Dick, 14, 85
Chambers, John, 15, 30, 31, 40, 41, **48**, 49, 55, 67, 73, 74, 79 110, 120, 124, 144
Commonwealth Antarctic Expedition, 131
Continental Motors Corporation, 68
Coulter, J.E. Ltd., 25
Coventry Climax Engines, **33**, 37
Crab four-wheel drive car, 147
Craven Wagon and Carriage Works, 33
Cripps, Sir Stafford, 66, 81
Curzon, Joe, **126**
Czarnocki, Witol, 80, **126**

D

D'Angelo, Horace, 78, 109, 121
David Brown & Sons Ltd, 30, 33-4
David Brown Tractors Ltd., 34, 39, 42
    Park Works tractor shop, **35**, **36**, 38
    VAK1 tractor, 42
Dick, Alick, **121**, 130, 144
Dixon, Freddie, 147
Dowdeswell, Dick, 78, 89, **126**, 127, 130, 137
Dowdeswell, Ted, **72**, **119**
Duncan, James, 119-20, 122, 123, 128-9, 141

E

Eros, 21, 23

F

FF developments, 149
Farm Implement and Machinery Review, 122
Fergus car, 21
Ferguson, Betty, **47**
Ferguson, Harry
    inventor of the Ferguson System, 10-15, **10**, **16**
    background and early career, 16-27, **19**
    alliance with David Brown, 28-43, **32**
    collaboration with the Ford Motor Company, 46-61, **47**, **48**, **50**
    developing the TE-20 with the Standard Motor Company, 64-93, **65**, **72**
    export business, 94-115, **108**
    deal with Massey-Harris, 119-29
    involvement with car design, 147-8
Ferguson, J.B. & Co., 19
Ferguson, James, 18
Ferguson, Joe, 18-19, 21
Ferguson, Mary, 18
Ferguson-Brown Ltd, 39, 42
Ferguson combines, **126**, **127**, **128**, **129**, **130**, **131**, 134-6
Ferguson Holdings Ltd, 66
Ferguson implements
    3-Ton trailer, 75, **105**, 106, 107
    30cwt trailer, **75**, **76**, 77, 93, **144**
    Blade terracer, 87, 130
    Buckrake, **90**, 93
    Cordwood saw, 75, **77**, 77
    Cutrake, 133
    Disc terracer, 133
    Earth-leveller, 87
    Earthmover, 92
    Earth scoop, 75, 77, 130
    Fork-lift attachment, 133, 142
    Game flusher, 93
    Hammer mill, 87
    Harrows
        Disc, 75, 77, 82, 102, 108
        Mounted offset disc, 92
        Reversible heavy-duty disc, **94**, **126**
        Spike-tooth, 75, 77
        Spring-tooth, 75, 77
        Tandem Disc, 56, 93, 134
    Hay sweep, 92
    Hedge-trimmer, 86, 133, 152
    High-lift loader, **86**, **88**, **89**, 92, 93, **122**, 130
    Hydrovane compressor, 92, 133, **152**
    Irrigation pump, 106
    Linkage winch, 87
    Manure loader, 69, 86
    Manure spreader, 69, 87
    Mowers, 75, 77, 108
        Cutterbar, 56
        FE-79 mid-mounted, **139**, 141
    Ploughs, 77, 87, 102, 108
        "Belfast", **21**, **22**, 23
        Disc, **100**, 102, 107, **119**
        Duplex, **16**, **22**, **23**, 24
        FE-93 mouldboard, 141
        Ferguson-Sherman, **24**, 25
        Reversible, 92
        Single-furrow, 42, 93

Two-furrow, 73, 75, 92
Three-furrow, 71
Type B two-furrow, 38
Polydisc seeder & combined cultivator, 133, 133
Post-hole digger, 75, 77
Potato planter, 86
Potato spinner, 56, 86
Precision seeder, 92
Rear-mounted crane, 133
Rice transplanter, 102
Ridgers, 75, 77
Type D, 38, 47
Rowcrop cultivators, 47, 75
Type E, 38
Rowcrop thinner, 132, 136, 137
Seed and fertiliser drill, 92
Side-delivery rake, 125, 133
Spinner broadcaster, 133, 140
Sprayers
Low-volume, 91, 93, 125
Medium-pressure, 133
Spring-tine cultivator, 56, 77
Steerage hoes, 72, 75, 77, 93, 134
Subsoiler, 87, 108
Sugar beet cultivator, 56, 59
Sugar beet harvester, 132, 133, 136-7
Sugar beet lifter, 134, 136
Sugar beet topper, 137
Tillers, 56, **72**, 75, 77, 102
Type C, 38
Transport box, 75, 77, 106, **140**
Weeder, 87
Winch, 130
Ferguson School of Farm Mechanisation, 14, **79**, 85
Ferguson-Sherman Incorporated, 25
Ferguson-Sherman Manufacturing Corporation, 50-1
Ploughs, 50
General-purpose cultivator, 50
Inter-row cultivator, 50
Ferguson system, 10-15, **12**, 50, 51, 66, 79, 93, 114, 120, 142
Ferguson tractors
"Black", **12**, **29**, 31, **31**, 43
FE-35, **105**, 108, **134**, **135**, **136**, **137**, 138-42, **138**, **139**, **140**, **141**, **142**, **143**, 161
FF-30, 104, 105
FF-30DS, **103**
FF-30GE, 104
Ferguson 40, **112**, 115
Hi-40, **113**, 115
Ferguson-Brown (Type A), 29-43, **29**, **31**, **32**, **33**, **34**, **35**, **36**, **37**, **38**, **39**, **40**, **41**, **43**, **46**, 160
TE-20, 11-14, **11**, **12**, **13**, 62-93, **63**, **65**, **67**, **68**, **69**, **70**, **71**, **72**, **80**, **81**, 96-110, **105**, 106, 123, 127, 129-30, 134-5, 141, **151**, **152**, **153**, **156**, 160, 161
TE-A20, **73**, **74**, 81, 82, 87, 93, **101**, 103, 104, **122**, **123**, **124**, 131, **158**, 161
TE-B20, 83, 161
TE-C20, **75**, 83, 104, 161
TE-D20, 75, **76**, **77**, **78**, 84, 88, **89**, **90**, **92**, 107, **117**, **154**, 161
TE-E20, 84, 161
TE-F20, **82**, **83**, **84**, 89-90, 93, **94**, **96**, 104, **121**, **125**, 152, **159**, 161
TE-G20, 104, 161
TE-H20, **79**, 86, **96**, 161
TE-J20, 86, 161
TE-K20, 91, **102**, 104, 161
TE-L20, 91, 161
TE-M20, 91, 161
TE-N20, 104, 161
TE-P20, **85**, 91, 161
TE-R20, 91, 161
TE-S20, 91, 161
TE-T20, **86**, **87**, 91, 161
TE-PT20, **86**, **87**, 161
TE-TT20, 161
TE-PZD20, 161
TE-PZE20, 161
TE-TZD20, 161
TE-TZE20, 161
TE-60 (also known as LTX), **119**, **120**, **121**, 122-9
TO-20, **108**, **109**, 110, 112, 160
TO-30, 110, **111**, 113, 160
TO-35, **111**, **112**, 113, 114, 129, 138, 161
Ferguson tractor conversions, 151-9
Bombardier half-tracks, **122**, **124**, 130, 131, 133
Borrow mobile drier, 159
Bryden Tracpac, 154-6, **154**, **155**
Full tracks, **123**
Lenfield high-clearance sprayer, 157-8, **157**, **158**
Perkins P3, 151-3, **151**, **152**, **153**
Reekie narrow ("Berry") tractor, 157, **156**, **157**
Roadless DG13 half-tracks, **155**
Selene four-wheel drive system, 156-7
Twose Tractamount roller, 159
Fisons Pest Control, 133
Fitchett, Morris, 41
Ford, Edsel, 50, 52, 58
Ford, Henry, 42, 46-53, **47**, **48**, **50**, 59
Ford, Henry 11, 59-60, 109
Ford Hydraulic System, 61
Ford Motor Company, 25, 26, 46, 49, 53, 55-61, 64, 113
Dagenham, 49, 53
Highland Park plant, 60
Rouge River plant, 48, 50
Ford tractors
8N, 61, **61**
NAA, 61
Ford Tractor Operations, 11
Ford Ferguson tractors
9N, 11, **44**, **48**, **49**, **50**, **51**, **52**, 48-57, 60, 103, 160
9NAN, 52, **53**, 54, **151**
2N, 53-5, **54**, **55**, **56**, **57**, **59**, 60, **60**, 61, 64, **153**, 160
2NAN, **60**
Moto-Tug, 55, **58**
4P, 57-8, **58**, **59**, 123
Fordson tractors, 11, 23, 24, 27, 39
E27N Major, 57, 82
Model F, **16**, **22**, 23, **24**, **25**, 26, 46, 52
Model N, 46, 53
Foxwell, John, 11
Fraser, Colin, 14
Fuchs, Vivian, 131

G

George Bryden Engineering, 154-6
Gladwell, Arthur, 42
Gladwell & Kell Ltd, 42
Goodenough, Len, 127
Greer, Archie, 15, 25, 30, 31, 34, 49, 55, 67-8, 74, 79
Greer, Thomas McGregor, **16**, 19, 38
Grinham, Ted, 80, 88

H

Halford, Bill, **126**, **127**
Harriman, Roy, **96**
Harrow, Bill, 125
Harry Ferguson de France, 104
Harry Ferguson Incorporated, 51, 57, 60, 66, 73, 78, 93, 109, 110, 112
  Ferguson Park, Detroit, **106**, **107**, **108**, 110, 143
Harry Ferguson (India) Ltd, 106, 107
  Madras plant, 106
Harry Ferguson Ltd, 20, 24, **26**, 26, 34, 38, 39, 66-7, 78, 79, 83, 85, 104, 110, 122
  Banner Lane, *see under* Standard Motor Company
  Fletchampstead, **31**, **66**, 67, 85, 120, 126, **129**, 131, **133**, **134**
  Packington Hall, 85
  Powerscourt, 85
Harry Ferguson Motors Ltd, 53, 55-6, 59
Harry Ferguson Research Ltd, 147-9
  Prototype cars, 147, **147**, **148**
  Project 99 racing car, 148-9, **148**
Heath, Roland, 106
Hennessy, Patrick, 25
Herald, Ralph, **79**
Hercules Corporation, 30
Hill, Claude, 147
Hill, Captain Duncan, 121
Hill, Walter, 25, 40
Hillary, Sir Edmund, 131
Hockey, Stan, 93, 135
Holden, Eric, **117**
Hotchkiss-Delahaye, Societe, 104
Howell, Horace, 136

I

Imperial Motors, 38
Industrija Motora Rakovica, 108
Industrija Motora I Traktora, 109
Industrija Poljoprivrednih Masina, 108
Industrija Traktora I Masina, 108
International tractors,
  Titan, **20**

J

Jensen Motors, 149
Johnson, Lieutenant-Colonel Philip, 27

K

Ketchell, Bill, 107
Klemm, Hermann, 73, 110, 111, 113, 128, 137, 138, 145
Knox, Trevor, 34, 64, 79, 120, 144
Kyes, Roger, 51, 66, 67, 78

L

Leigh, Lord, **139**
Lend-Lease agreement, 53
Lenfield Engineering, 157-8
Liney, Nan, **76**
Liney, Nigel, 78, 80, **81**, 89, 127, **137**

M

Massey Ferguson Ltd, 107, 115, 142-6
Massey Ferguson Museum, **35**, **124**
Massey Ferguson implements
  Plough, **105**
  Side-mounted baler, 115
  Forage harvester, 115
Massey Ferguson tractors
  MF-35, 142, **144**, 146
  MF-50, 115, **115**
  MF-65, 115, **115**, 142, **144**, 146
  MF-85, 115, **115**
  MF-835, 105
Massey-Harris 701 baler, **117**
Massey-Harris combines
  MH-735, 135, 136, **143**
  MH-780, **117**, 136, **143**
Massey-Harris Company, 93, 113
Massey-Harris tractors
  MH-50, **114**, 115, **121**
  MH-55K, **119**, **143**
  MH-744D, 118, **118**, 128
  MH-745, 118, **119**, 128, **143**
  MH-820 (Pony), 125
Massey-Harris-Ferguson Ltd., 93, 108, 113, 118-23, 129, 133, 141-2
Massey-Harris-Ferguson, Compagnie, 104
May Street Motors, 20, 22
McNeice, James, 121
Metcalfe, Geoff, **126**, 135
Milivojevic, Svetislav, 108
Morris Motors, 27
Moss, Stirling, 149
Muir-Hill Service Equipment Ltd, 25
Mulholland, Pat, 118

N

Nanda, Harry, 106, 107
National Farm Youth Foundation, 52
National Institute of Agricultural Engineering, 136
Nehru, Pandit, 106
Newbold, Nibby, 127, **135**
Newsome, Noel, 13
Nicholson, Charlie, **118**, **126**
Nuffield Organisation, 149
Nunn, Bill, 118

O

Overtime tractors,
  Model N, **20**

P

Patterson, Alex, 15, 55, 67, 74, 83, 120, 126
Perkins, Frank, Ltd, 108, 143, 144, 151
  TE-20, Perkins conversion, 150-3, **151**, **152**, **153**
Pierce, Les, 135

R

Ransome and Rapier, 27
Reekie Engineering, 156-7
  "Berry" tractor, **156**
Reid, Roger, **118**
Ricardo, Sir Harry, 138
Roadless Traction, 27
  DG13 half-tracks, **155**, 156
Robert Eden company, 157
Rolt, Stuart, 149
Rolt, Tony, **147**, 147
Rover Company, 27
Russell, Jim, 41

S

Sanders, Arthur Freeman, 88
Sands, Willie, 15, **16**, 22-5, 30, 31, 34, **48**, 49, 55, 67-8, 74, 79
Sargeant, Ron, 92
Saunderson tractor, **11**
Searle, Colonel Frank, 27
Selassie, Haile, 100
Selene company, 156
Senkowski, Alex, 75, 79-80, 83, 124, 148, 149
Sherman, Eber, 25, 46, **47**, **48**
Sherman, George, 25

Sherwen, Theo, 93
Société de Motoculture du Gard, 158
Sorensen, Charles, 23, 48, 49, 59
Standard Motor Company, 64-8, 74-5, 80-1, 88, 102, 104, 113, 121-2, 130, 142, 144-6
Banner Lane, 13, 63-6, **64**, **70**, **71**, 74-5, **82**, 96, **101**, **130**, **138**, 140-1, 143, 145-6, 147
Beauvais, 105, 146
Canley, **133**, 146
SMC prototype tractor, 145, **145**, **146**
Société Standard-Hotchkiss, 104, 105
St. Denis, **102**, **103**, 104, 105, 146
Standard Motor Products of India Ltd, 105
Standard 6cwt van, 93
Standard 12cwt van, **74**, **118**
Vanguard car, 81, **84**, 90
Staude, E.G., Manufacturing Company, 23
Steventon, Colin, **72**, 78, **118**, 127, **132**, 137
Stoltenberg-Blom, Jean, 126
Stoltenberg-Blom, Michael, 134

T

Thornbrough, Albert, 112, 129, 141-3
Thorne, Mike, **139**
Tractor and Farm Equipment, 107
Turner-Hughes, Charles, 78
Twose, E.V. Ltd, 159
Tractamount roller, **159**

V

Vincent, Charles, 66, 78

W

Wallace, Ian, 79
Waterloo Boy tractors, 22
Watson, Maureen, 19-20
White, Bud, 108
Williams, John, 19, 46, **48**
Williams, Les, **67**
Wilson, John, 79
Wright, Colin, 128, 130

Y

Young, Eric, 79, 120, 121

Organisations

The following two British clubs cater for the Ferguson enthusiast and both are useful sources of information.

FERGUSON CLUB

Chairman
Harry Turkington
37 Trench Road
Hillsborough
Co. Down
BT26 6JL

02892 688252

Membership
Lawrence Jamieson
PO Box 20
Golspie
Sutherland
KW10 6TE

01408 633108

FRIENDS OF FERGUSON HERITAGE

Membership
7A Church Way
Whittlebury
Towcester
Northants
NN12 8XS

01327 857426

Merchandise
PO Box 62
Banner Lane
Coventry
CV4 9GF

02476 852204

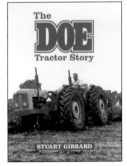

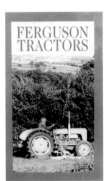

About the Author

A successful author and journalist specialising in tractors and machinery, Stuart Gibbard comes from a farming background near Spalding in Lincolnshire. He has developed his interest in collecting early tractor literature into a mail-order business which is run by his wife Sue, and he is also one of the organisers of the annual Spalding model-tractor and literature show.

Devoting much of his time to historical research, Stuart has talked to many of the men who played their part in creating the machines portrayed in his books. This first-hand knowledge has enabled him to give a fascinating insight into the world of agricultural engineering and tractor development, and his prize-winning publications include unrivalled histories of County, Roadless and other Ford tractor conversions.

Stuart is currently editor of *Old Tractor magazine.*

Publications

BOOKS

Change on the Land
County: a Pictorial review
David Brown Tractor Story Part 1
Doe Tractor Story
Ferguson Tractor Story
Ford Tractor Conversions
Ford Tractor Story Parts 1 & 2
Roadless Tractors
Tractors at Work Volumes 1 & 2
Tractors in Britain

DVDs

County Tractors
County Tractor Working Days
Ferguson Tractors
Ferguson on the Farm (with Harold Beer) Parts 1, 2 & 3
Fordson Farming
Giants of the Field
International at Doncaster

Fer